D0934956

Resolving the Innovation Paradox

Resolving the Innovation Paradox

Enhancing Growth in Technology Companies

Georges Haour

First published 2004 by
PALGRAVE MACMILLAN
Houndmills, Basingstoke, Hampshire RG21 6XS and
175 Fifth Avenue, New York, N. Y. 10010
Companies and representatives throughout the world

PALGRAVE MACMILLAN is the global academic imprint of the Palgrave
Macmillan division of St. Martin's Press, LLC and of Palgrave Macmillan Ltd.
Macmillan® is a registered trademark in the United States, United Kingdom
and other countries. Palgrave is a registered trademark in the European
Union and other countries.

ISBN 1–4039–1654–3

This book is printed on paper suitable for recycling and made from fully
managed and sustained forest sources.

A catalogue record for this book is available from the British Library.

Library of Congress Cataloging-in-Publication Data
Haour, Georges, 1943–
    Resolving the innovation paradox : enhancing growth in technology
companies / by Georges Haour.
        p. cm.
    Includes bibliographical references and index.
    ISBN 1–4039–1654–3 (cloth)
      1. High technology industries—Management. 2. Technological innovations—
Management. 3. Research, Industrial—Management. 4. Corporations—Growth.
I. Title: Enhancing growth in technology companies. II. Title.

HD 62.37.H36 2004
658.5′14—dc22   2003062318

10 9 8 7 6 5 4 3 2 1
13 12 11 10 09 08 07 06 05 04

Printed and bound in Great Britain by
Creative Print & Design (Wales), Ebbw Vale

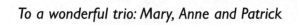

To a wonderful trio: Mary, Anne and Patrick

CONTENTS

LIST OF FIGURES AND TABLES

Figures

Table

This book concerns the practice of innovation. It is based on twenty-five years of experience in managing the innovation process and then envisaging that process from a more reflective vantage point. I hope that the experiences and ideas shared in this book will provide a stimulating contribution to managers, as well as to students of innovation in technology companies.

Humankind has demonstrated an extraordinary ingenuity in converting scientific and technical knowledge into useful artifacts. The combustion engine, airplanes, radio, the automobile, synthetic materials, pharmaceutical and medical technology, the personal computer, among others, have dramatically changed our lives over three generations. The internet, mobile telephony and biotechnology relentlessly impact our societies. Advances in these areas will result in a continued proliferation of new devices, therapeutic drugs and businesses, with varying degrees of success.

Science and technology are a crucial component of the human adventure, alongside the arts and literature. The eclectic genius of Leonardo da Vinci uniquely united these talents in one single person. At the business level, all stakeholders agree on the importance of technical innovations for the survival and growth of firms, yet it is rare for top management truly to focus on this priority. This paradox is all the more surprising given that it is increasingly difficult to secure and maintain high returns on investments in innovation. Companies must resolve this paradox with a new view of the idea-to-market process. A growing sense of urgency has prompted me to present such a view in this book.

The new framework of *distributed innovation* is a tool in the hands of top management for envisaging and steering innovation towards the strengthening of the competitiveness of companies. Aspects of this novel approach may well be applicable to companies in general, but I focus on technology companies because, more than ever, they represent the key to creating wealth and jobs in the future.

My whole professional trajectory has dealt with technology-based innovation, venturing and entrepreneurship in international settings. I have practised in the world of commercializing technology from the perspective of researcher, manager, management professor, venture coach and investor, and in this book I distill some of the lessons learned in order to provide a few beacons in our exciting, but sometimes confusing, times.

I am deeply indebted to many managers from all over the world, as well as to several colleagues, who took the time to discuss issues of common interest in the field of management, where there is no panacea and where each situation is unique. I thank them all warmly for their generosity of spirit in sharing their experiences and wisdom with me.

It was indeed a pleasure to work with Stephen Rutt of Palgrave Macmillan in the course of producing the manuscript.

Geneva and Cambridge   GEORGES HAOUR

CHAPTER 1

Innovation is Survival

Innovation is central to the wellbeing of societies, as well as to the health and growth of commercial companies. It represents a great leverage in creating economic value. The penalty for not innovating is enormous. Innovation manifests itself in many different ways and is very hazardous to predict, both in its timing and in its consequences. It is difficult to manage the process of making it emerge and succeed.

The last decades have seen an enormous generation of technical knowledge. The pace of change in the societal and business environments has been unprecedented. This should make the striving for innovation, and technical innovation in particular, a top priority on the agenda of countries and companies. The paradox is that this is often not the case. As a result, the flow of needed innovations is far from optimal. Is there now an innovation crisis? Transformational innovations are needed more than ever and this book proposes an approach to respond to this need.

Innovate or Evaporate

Innovation is the life-blood of competitiveness. A few years ago, Singapore started a campaign aimed to foster innovation in the city-state, primarily, but not exclusively, in the area of technology companies. It came up with the slogan 'innovate or evaporate', which I have borrowed as a title for this section.

Innovation is invention converted into a product, an industrial process or a service for the marketplace. The cellular phone, continuous casting of steel, internet banking, or the self-service store, are examples of innovations. Innovation may also be a new way of doing business, and examples include easyJet in the airline industry and the ill-fated Enron for energy trading. As shown by these examples, innovation is much broader than just technology in its nature.

Effective innovation represents *the* way for companies to escape the downward spiral of diminishing returns which comes from relying only on operational efficiency. Schumpeter's phrase 'the gales of creative destruction' puts innovation and entrepreneurial energy at the centre of renewal and economic growth. Schumpeter himself did not use the word innovation, but he was very close to the definition above when he described the economic benefits derived from 'the introduction of a new good, the introduction of a new method of production, the opening of a new market, the conquest of a new source of supply of raw materials or semi-manufactured goods, and the carrying out of a new organization of any industry, such as the creation and break-up of a monopoly'.[1]

'The gales of creative destruction' take corporations by storm. Many disappear as a result: who remembers Digital, the electronics industry leader in the 1980s? Other firms leverage change for profit and growth. Societies demonstrate varying levels of acceptance of such changes. In the 1970s, Japan enjoyed change, as it produced strong economic uplift; two decades later, the same country is lastingly bogged down for denying the need for change while many companies in Japan are continuing to innovate at full capacity. The start-up companies are the embodiment of the power of innovation: Silicon Valley and other similar regions in the world have caught the public imagination because they offer exciting examples of Schumpeter's insight.

In these regions, innovations based on technical development represent the main fuel powering the emergence of new firms. In this book, I concentrate on innovation in technology companies, because this segment is a key source of value creation. These firms aim to convert technical expertise into products and services for the marketplace, which are either to be sold to individual end consumers – as is the case for consumer electronics Sony – or to be sold to another firm in a business-to-business mode – airplanes from Airbus, for example.

The term 'technology companies' refers to firms that internally generate a substantial amount of technical knowledge, primarily in their Research and Development (R&D) functions. These include companies in sectors such as computers, software and telecommunications, pharmaceuticals and biotechnology, medical equipment, specialty chemicals and materials. In contrast, service companies such as airlines, banks, insurance firms and retail businesses are primarily *users* of technology. Airlines, for example, use technologies, particularly information and aircraft, developed and manufactured by technology companies such as Airbus or Boeing, Bombardier or Embraer.

When looking at R&D investments, there is no conclusive evidence of a correlation between them and the subsequent financial performance of the firm over time. Similarly, it is not clear whether over time the financial markets truly reward companies which invest in R&D in a sustained way. However, in any given industrial sector, what is clear is that technology companies grow faster when they invest more heavily in R&D. Furthermore, the penalty for not preparing for the future is clear. Still, many firms would like to know what level of investments is necessary to remain competitive. No such luck! Crucially, as we will see below, this issue is far from being purely a matter of investment figures.

In a given industrial sector, companies invest in R&D roughly the same percentage of sales. For example, this number is in the 16–20 per cent range for pharmaceutical companies. There are, of course, differences from company to company; such variations may, however, be a result of using different ways of computing the statistics. They can also result from the different ways in which technology firms carry out their innovation process.

Putting Technological Innovation to Work

While in technology firms innovation is not the sole property of technologists, the R&D function is a key actor in that effort. Going from idea to market mobilizes the whole company. Harnessing technology for business advantage is a kind of Odyssey, in which the ship of innovation sails on tenaciously, resisting the sirens singing of false breakthroughs, and fighting the monstrous Cyclops of the NIH – Not Invented Here – syndrome.

Today's pool of technical know-how is larger than ever. It is estimated that close to four million professionals are involved in

R&D worldwide.[2] More than 80 per cent of them live in the first world, primarily in the United States (40 per cent), the fifteen countries of the European Union (26 per cent) and Japan (15 per cent). In the 29 richest countries grouped in the OECD, the average ratio of researchers per thousand people in the work force is now 6.2, as compared to 5.6 in 1990. This underscores the development of what is often called the 'knowledge economy'. The United States, Sweden and Japan have the highest ratios, in the range of eight to nine researchers per thousand workers.

These professionals build on the 'shoulders of giants', constituted by the results of previous efforts, which have grown exponentially in the last decades. In the member countries of the OECD, investments in R&D by public and private sources, have roughly doubled in 20 years, to reach $600 billion in 2000.[3] The bulk of this growth comes from the industrial sector. In 2000, the share of total R&D funded by industry was 64 per cent in these countries. Such input numbers, however, constitute a very imperfect measure for our present purpose. The measure of interest to us is the return on these investments. In successfully commercializing technology, the key is the quality of output. In the end, this is measured by the business success and growth of technology companies. No single indicator, whether it is the number of patents granted or any other statistics, truly reflects the productivity of converting science and technology into profits and growth.

Part of the challenge of knowing how effectively technological innovation is converted into commercial success comes from the difficulty of calculating the cost of the innovation investments *a priori*, and often even in hindsight, when the product has come to the end of its life in the marketplace. This difficulty is a result of several factors. First, as already mentioned, R&D is only one of the actors in the innovation process. What is also needed is to be able to calculate the total investments, not only in R&D, but also in marketing, design, manufacturing, legal work and, in particular, general management time. Second, even just R&D investments are difficult to clearly relate to any specific product or service in the marketplace. The 'critical path' of R&D investments concerned with development runs through many different projects over several years, probably carried out in several organizational structures. Additionally, unrelated or failed projects may well have contributed critical knowledge to the development in question. Such 'spill over' benefit is extremely problematical to trace reliably. The above input numbers, however, demonstrate that technical knowledge is increasing at a considerable rate. With regard to investments in R&D activities, it is customary to distinguish between basic research, applied research and experimental development, as in the OECD Frascati manual.[4] For our purpose, however, and following recent work,[5] we prefer to distinguish two broad classes of R&D activities.

R&D Motivated by Curiosity

This activity is mostly financed by public funds and is primarily carried out at universities and government laboratories. It is not targeted at specific commercial applications and also provides the more independent and objective component of scientific and engineering pursuits. This type of longer-term research is where developments begin, which many years later may have very important commercial consequences. Laser technology is such an example. In the 1970s, Nobel prizes were granted for work in this area, although at the time industry was uncertain what might be their viable applications. Thirty years later lasers have a very wide range of applications, from covering metrology and measurements to treatment of materials and defence.

In contrast, the invention of the transistor in 1947 quickly led to an immediate application: Sony developed pocket-sized radio sets – 'transistors' – in the 1950s. The Human Genome Project is another example. Financed with public funds, this groundbreaking work will find practical applications with a time horizon in the order of a decade, in areas such as diagnostics and therapeutic treatments.

This type of R&D is one of the best investments a country can make for its future. It is relatively inexpensive: it represents less than 1 per cent of the GNP in industrialized countries and generates handsome returns in terms of real value creation for the country, as well as productivity improvements for its industry. This kind of investment is typically at the origin of new sectors of technology-based activities, such as the ICT (information, computer and telecommunications) industry or the genomics/proteomics sector. In the 1990s, the United States was one of the few countries in the world that increased its efforts in this area, while countries like Japan and Sweden maintained their high level of efforts.

Another example of the impact of publicly funded R&D is the invention of the architecture of the Web at CERN – European Centre for Particle Physics. This centre is supported by a large number of governments worldwide. In the process of improving ways to connect many hundred professionals working on an experiment on its particle accelerators, CERN came up with the concept of the World Wide Web, which was then further developed in the USA, partly with Defence funds. It is interesting to note that a need for communications of a research laboratory was part of the development of such an important phenomenon as the Internet.

Results obtained in the course of publicly funded, curiosity-driven research are generally available and published in scientific journals. This evokes the phrase 'Science as an open house', coined by Robert Oppenheimer, the most senior physicist of the 'Manhattan Project' for the development of the nuclear bomb in the USA in the 1940s. Increasingly, the object of scientific output is filing patents, so that the government or university laboratories carrying out the research will be in a position to commercialize its results later. In many ways, curiosity-driven research, sometimes wrongly called 'pre-competitive research', is coming closer to some of the characteristics of business-motivated R&D.

R&D Motivated by Commercial Objectives

Business-driven R&D aims at bringing specific products or services to the marketplace. It deals with developing products as well as ways of manufacturing them. It is overwhelmingly financed by private companies, but public support is sometimes available for this purpose, particularly in areas such as life-sciences and defence. Financing may also be shared on a private/public basis of 'matching funds', but the bulk of this type of R&D is carried out by companies. It may also be conducted by other organizations, such as universities or specialized laboratories on behalf of industrial clients. This book concentrates on innovations supported by this type of R&D.

Such innovations may be long term – the development of a new drug can take ten to fourteen years from the discovery of the molecule to market launch – for most innovations, the time horizon is typically mid- to short-term. For example, it took one to three years to develop the 'Walkman', a new microwave oven, a new car model, some software. It becomes much shorter – only a few months – if it involves a slight modification of an existing product or process. The term 'incremental innovation' is used in this case.

Distinguishing between curiosity- and business-driven motivations is preferable to the previous dichotomy between 'basic' and 'applied' types of R&D because it better reflects the difference in funding, and also because the boundaries between 'basic' and 'applied' are increasingly blurred, as for instances in the case with the micro lithography process of producing microchips. This industrial process affects the matter at the atomic scale, so is work towards its improvement of a basic or applied nature? Other examples are nanotechnologies, genomics and proteomics, where scientific knowledge is deployed in advanced technologies, in commercial pursuits to identify new molecules and therapeutic approaches in the life-sciences sector.

The enormous pool of scientific and technical knowledge already accumulated is growing fast. It provides an unprecedented source for the firm to draw on. The key is to do so effectively. Innovation is indeed primarily a matter of *effectiveness*. In developing technical innovations, how well are we doing in this respect?

An Innovation Crisis?

During the last two decades, two countries have successively served as models for innovation: Japan and the USA. They followed very different approaches.

In the 1980s, the world was fascinated by Japan's prowess in combining excellent engineering skills with keen business sense and relentless customer orientation. This was particularly true in the steel, automotive and consumer electronics industries, with star companies such as Nippon Steel, Honda and Toyota, Canon, Hitachi, NEC, Matsushita and Sony. Companies in the West learned many lessons from their Japanese counterparts and tried to apply them to their own operations. The world watched in awe as Japan elevated its status from that of an emerging nation to the second world economic power in the space of just 20 years (the first *shinkansen* high speed train project was completed in 1964 with World Bank financing). Japan also amazed the world with the extensive cooperation that took place among competing technology companies. The role of the ministry of industry, MITI (now METI), as a *deus ex machina* of the country's economic development, was the object of much debate. Now, in the early 2000s, Japan is perceived to have 'lost its touch', an assessment as extreme and incorrect as in the previous period of fascination.

In the early 1990s, the United States became the role model, surfing on the dynamics of innovation and entrepreneurship during a decade of growth. The energy and talent of the dense region of Silicon Valley, south of San Francisco, powered a remarkable wealth-creating machine. Large companies were born in that small area: Hewlett Packard, Intel, Oracle, Sun. The industrial biodiversity of the region included hundreds of fast-growing technology start-up companies, mostly on the Internet, but also in life-sciences. These companies were chronicled by the magazine *Red Herring*, now defunct, with a taste for technology and money. The collapse of the stock market in the Spring of 2000 prompted healthy questions on the way all the actors involved – analysts, banks, media, auditors, consultants, managers – conspired to reinforce each other in a naively optimistic view that this was a different world, with only the sky as the limit. The bursting of the bubble, with the corresponding precipitous drop in venture capital activities, compounded an ongoing 'innovation productivity crisis'.

Part of the 'innovation crisis' comes from the entrenched model of internal innovation, which emerged after World War II. Consider this question: as companies grow, by mergers or acquisitions, should their R&D investments continue to represent the same fraction of the larger sales? Or should there be some kind of economy of scale? If so, this would mean that the firm, having acquired more muscle in distribution and sales, should not allow its investments in innovation to grow at the same rate.

This let-up is observed in technology start-up companies. In the early years, technology start-ups must finance major innovation projects. As sales take off, the unsustainably high R&D investments proportionally decrease. However, in large and established companies, what is generally observed is as follows: as the company sales increase, it still keeps the R&D investment at the same percentage of the turnover in order to provide insurance for the future business.

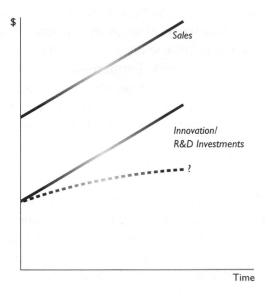

Figure 1.1 Innovation/R&D investments in a company (as its sales volume grows, should the firm continue to keep the same percentage of sales (full line) or should economies of scale (dotted line) allow a reduction of that ratio?)

This is illustrated in Figure 1.1. It looks as if, by a leap of faith, the company increases its investments in innovation activities proportionally. Could they invest slightly less and still achieve the same growth?

This in turn raises the central question: when starting from scratch, how much should a firm invest in R&D? The difficulty in answering that question leads each company to invest more or less at the same level as the rest of the industry. Here again, however, what really counts is the quality of the output. One would therefore expect managers of technology firms to be passionate about the effectiveness of their innovation project teams, since their talent and motivation have a major impact on the quality of the outcome. The paradox is that this is not a high priority in most technology companies.

The 'innovation crisis' is illustrated by the pharmaceutical industry. As we have seen, that industry invests on average 13 per cent of sales in R&D. In 2001, this represented $45 billion (27 billion in the USA alone) for that industry worldwide.[6] In spite of increasing investments, the number of new drugs, the NCEs – New Chemical Entities – resulting from these efforts has been decreasing since 1987 from 25 to 20 per year in Europe, while slightly increasing in the

USA from 13 to 14 per year. As a result, the cost per NCE has grown tremendously in recent years, to reach an estimated $800 million in 2002,[7] because this average also has to take into account the costs of the unsuccessful developments. Furthermore, out of all new molecules introduced each year, only a quarter of them generate revenues in excess of the developments cost.[8] The overall effective productivity of the drug development process has thus declined dramatically. The new approach for drug discoveries has yet to prove effective in reversing this trend. This involves bioinformatics, combinatorial chemistry and simulation, with corresponding high technology investments of more than $100 million per company per year.

As indicated by the example of the pharmaceutical industry, our current innovation model seems to be reaching a limit. The law of diminishing returns compels us to look for an alternative model. The aim of that new model must be to better take into account the new realities of technical developments in our times. This means a break from the attitudes established in the last decades. Currently, the innovation process is essentially internal to each company, with only *ad hoc* external collaborations carried out in a piecemeal way. This book proposes *distributed innovation* as a novel approach for technology companies.

Overview of the Book

Distributed innovation aims to leverage technical expertise more effectively. This book presents a number of insights and suggestions for applying this new perspective to innovation in technology companies.

Chapter 2 focuses on the *innovation paradox*, showing that, despite the acknowledged critical importance of innovation, companies' top management, the CEO – Chief Executive Officer, does not in fact place this issue at the centre of the radar screen. This is a result of short-term financial pressures on the CEO. Further, in addition to absorbing much attention and energy, the tyranny of constantly working to make sure that the firm is *perceived* positively by the financial community creates a mindset that detracts from building and nurturing a culture of risk-taking through innovation, especially for the longer term.

Hence the *innovation paradox*: although recognized as an absolutely crucial ingredient for business growth and profitability, innovation does not receive the priority treatment it deserves from top management. Innovation is not a matter of efficiency, but of *effectiveness*: strong commitment to it is required from every level, starting at the top. There is a pressing need to correct the current benign neglect and to swing the pendulum away from the urgent pressures of short-term financial results towards innovation-led, profitable growth.

Chapter 3 explores the technology management practices used in steering the innovation process. Currently, firms rely essentially on internal innovations, complemented by *ad hoc* collaborations with external organizations. Tools and practices are described, which companies put to work in an attempt to improve the business impact and market-relevance of this process. The impetus for such changes came as a response to Japan's economic success in the 1980s.

Chapter 4 describes the unique business model of the company Generics of Cambridge, UK. This innovator/incubator/investor company creates value out of technical expertise in many different ways. It is proposed that technology firms should follow this example by being more adept at creating value by proactively using various channels of commercializing technology.

Chapters 5 and 6 describe the array of various actors that constitute the *distributed innovation system*. In order effectively to steer innovation for growth in today's world, it is argued that it is not enough to have a vital and productive internal innovation process. *Distributed innovation* extends the company's innovation perimeter far beyond the boundaries of the company. In this way, the options available to the firm are greatly expanded. This new approach to innovation involves managing the complexity of dealing with different technology channels, as well as interacting with various external actors.

Chapter 5 explores one application of this approach. It aims proactively to convert technical expertise into new revenues for the company, using different channels to commercialize technology. These include licensing, selling innovation projects and innovation mining. It demands that the firm demonstrate a great awareness of the business and technical environment outside the firm.

Chapter 6 explores how these same external actors are also potential sources of technology which could be tapped by the company. Distributed innovation is an innovation-led approach carried out with an entrepreneurial perspective. The firm *sees* opportunities in the marketplace and marshals the necessary resources to address the opportunities it has selected.

The company thus defines specifically selected 'high impact' products or services. In mobilizing the technical resources necessary to develop them, the firm draws extensively on external actors to raise their technology, complementing its own internal capabilities. External and internal inputs are used 'seamlessly' for effective development. In defining and undertaking these specific development projects, the CEO must be seen as taking risks and should demonstrate full commitment to making them a success.

Chapter 7 stresses the crucial importance of fostering high levels of motivation among the team members of innovation projects. The quality of output of such projects is very dependent on this human factor. In carrying out development projects, the practice of *distributed innovation* requires a high level of trust within the firm. It also injects a strong outward perspective into the whole company, and the R&D function in particular. The resulting keen awareness of the external environment is most beneficial to the company in today's business world.

Chapter 8 concludes with *distributed innovation* as a tool available to the CEO to resolve the innovation paradox. This, it is argued, will enable the harnessing of technical expertise for maximum growth and profit.

The introduction of a *distributed innovation* system involves complex management challenges, but can provide handsome benefits in achieving more effective value-creation through technological innovation.

If they implement this new approach, technology firms will improve their entrepreneurial perspective, and will act more and more as *architects* of innovation, to some extent at the expense of the internal component of innovation management. The R&D function will thus extend to become an active *broker* of technology. More than ever, it must retain its technical edge. As for the CTO – Chief Technology Officer – he or she will increasingly be in charge of the business development of the company.

Notes

1. Joseph A. Schumpeter, *The Theory of Economic Development* (Harvard University Press, Cambridge, MA, 1911).

2. OECD *Science, Technology and Industry Outlook* (OECD, Paris, 2002). The site is: www.oecd.org
3. Ibid.
4. *Frascati Manual* (OECD, 1994).
5. *Encyclopédie du Management* (Dalloz, Paris, 1999), pp. 1034–7.
6. www.csdd.tufts.edu
7. *Ibid.*
8. *Ibid.*

CHAPTER 2

The CEO as Innovation Champion

Considering the paradox that although innovation is central to the competitiveness of firms, CEOs – Chief Executive Officers – of companies rarely put this issue on top of their action-list, resolving it involves providing appropriate incentives to the CEOs, while fostering a governance system more supportive of value-creating innovations over the longer term.

'Today's innovations create tomorrow's jobs' is a common phrase. In this regard, annual reports of companies make repetitious reading for they contain remarkably similar statements, as if they came out of the same corporate communications boutique. In these reports, standard phrases speak about customers' proximity, shareholder value, sustainable development, and so on. One such statement refers to the vital importance of innovation for the future growth of firms. Everyone seems to agree on the crucial role of innovation for growth and value-creation. The paradox is that, in reality, the actions of relatively few CEOs demonstrate true commitment to the issue.

The current system constitutes an environment that often does not promote innovation-led growth. In companies with a traditional shareholder-ownership, CEOs do not feel truly encouraged to leverage the power of innovation for growth over the longer term. There are exceptions in that some of them have the courage to be 'innovation activists' in their own firms.

Private or family-owned companies operate somewhat differently in this regard. Looking at this other type of governance, there may be ways to gently swing the pendulum towards making the innovation process more effective for the firms' longer term.

Does the Current System Encourage Innovation-led Growth?

In large technology companies, many elements detract from having a sharp focus on innovation. Many ambassadorial tasks shape the mindset and agenda of the CEOs. CEOs are caught between the tyranny of short-term financial results and building the business for the longer term through patient investments in innovation. There is a permanent tension between the urgent and the important. True leadership in favour of innovation is rare.

Elements detracting from committing to innovation are well illustrated by several examples from very large technology companies. The $126 billion North American technology conglomerate General Electric was founded by the archetype of technology entrepreneurs, Thomas Edison, for whom the business principle was: 'Anything that won't sell, I don't want to invent'. Yet, in recent years, a growing fraction of GE's profits has been generated by the non-technical component of its activities, the financial services division of GE Capital. According to its 2001 annual report, that percentage was a quarter of the operating profits. This paradox does not build much of a case for firms to put technical innovation at the centre of their radar screens.

In the pharmaceutical industry, celebrated as 'R&D driven', companies in fact spend, more on marketing and sales than they invest in R&D. Furthermore, they secure a large part of their profits not so much from their industrial activities as from financial investments in portfolios of stocks. This diversification boomerangs when the stock prices drop. In February 2003, the Swiss pharmaceutical company Roche announced an exceptional 5.1 billion Swiss Francs financial charge for the decrease in value of its portfolio of shares.[1] It is paradoxical that a so-called 'R&D-driven industry' ultimately owes so much of its performance to a portfolio of investments, rather than its sales of drugs and the robustness of its innovation pipeline.

Similarly, the German technology giant Siemens, with a yearly sales volume of Euro 84 billion, is often described thus in its home-city of Munich: 'Siemens is a financial institution, with some industrial interests'. This sounds like a paradox for a conglomerate which invests Euro 6 billion per year in R&D.

These examples suggest that other elements of their operations turn technology companies away from focusing on innovation as the key source for profit and growth. The emphasis on innovation to create value is distracted by the non-industrial activities of technology companies and the need to manage for high financial performance in the short term is prioritized over building for the longer term.

Putting the power of innovating into practice requires that the top management has a builder's mindset – and enough time. According to a recent study,[2] the average tenure of a CEO in an North American company is six years; the longest is ten years for firms in the financial sector, with only four years in the currently troubled telecommunications sector. With such a brisk rate of turnover, whatever points towards building a company's business can effectively be put into place, especially, as is often the case, if the CEO comes from outside the firm and thus needs time to learn the environment?

A somewhat exaggerated, but plausible, account of the tenure of a CEO may be the following: in the course of the first year on the job, the new CEO brings his own trusted people into his entourage, changes the company's organization chart, launches restructuring programmes for cost savings and concludes a few mergers and acquisitions deals. For all practical purposes, the CEO has a free ride during the first year of his tenure. During this 'honeymoon period', it is difficult for the board members to challenge the CEO, since this would appear to indicate that they recant on their choice. Furthermore, an effective critique of the CEO's decisions by raising tough business questions requires serious homework on the company and its environment. As a rule, board members do not have the time to carry out such homework, as they are busy with their other obligations. The recent debacles of AOL-Warner, Swissair and Vivendi, to name but a few, illustrate this problematic aspect of the 'governance' of companies, not to mention frauds like Enron, Worldcom and Daewoo, the Chairman of which is in hiding.

In his second and third years, the CEO focuses on consolidating the financial returns of his cost-cutting measures and on postponing longer term investments in order to boost shareholder value in the short term. Somewhere along the way, however, unpredictable and drastic changes in the external environment are likely negatively to affect the profitability of the company, so that shareholders, pension funds and others, become anxious at no longer seeing their profit range of 8 to 12 per cent. The CEO then loses support of the board, spends much energy on board politics and damage control, but soon counts his/her blessings for having negotiated a 'golden handshake' as indemnity for departure.

While employed, the CEO's mindset and priorities are centred on working to have the firm *perceived* positively. This includes convincing the analysts, participating in numerous, appropriately named, 'road-shows'. Time is also spent engaging and convincing the trade press and the general media. In this exercise of seduction of the media, CEOs find themselves in the same league as politicians and entertainment personalities. Many of them enjoy the attention of the media and become masters at 'communicating' a carefully crafted image. Corporate communications reinforce a positive image of the CEO. Media seem to be particularly keen on two types: on the one hand there are the extrovert, flamboyant CEOs, such as Virgin's Richard Branson, CEOs of many Silicon Valley companies or Jean-Marie Messier, former CEO of Vivendi, who once said 'Do not ask a CEO to be modest'. On the other hand, there are serious, cocktail-shy personalities like Percy Barnevik. The perceived personality of the CEO may temporarily contribute to the 'brand equity' of the firm.

The CEOs must also deal with the politics of the board in order to maintain the necessary level of support. Once the urgent ambassadorial tasks are discharged, attention is given to operations and staff, but by that point not much is left for laying the foundations of a solid longer-term future of the firm. Nowhere at the top of the CEO's agenda is working hard to make the innovation process work more effectively and being attentive to turning the portfolio of innovation projects into a healthy longer term competitiveness of the firm. Within the company itself, reasons for this are numerous. It is difficult to master innovation development and to predict its outcome. Technological issues are intricate and it is tempting to neglect the 'black box' of technology. Furthermore, the technical staff does not make enough effort to explain the implications of this 'box' and to bridge the gap between technology and business. This frustrating situation sometimes leads the CEO to entrust the responsibility of product development to technical specialists, thus further widening the gap.

In times of economic uncertainty, which represent on average half of the total time in any given region of the world, this predicament is compounded: corporations cut everything that has a short-term positive impact on the balance sheet and a negative one in the longer term. The items traditionally cut are: travelling expenses, use of external services, as well as innovation projects for the mid-term, when, in fact, this would be the time to invest counter-cyclically in innovation in order to come out stronger at the end of the bottom of the cycle. During such periods, financial pressures also require getting rid of equity positions in firms having temporarily low value, just to get them off the balance sheet, when, in fact, the best long-term interest of the company would be to retain these shares for better days. In this way, many companies, such as Compaq and Lucent, divested their corporate venture activities at substantial losses in 2002. This destruction of value for the firm is forced by short-term financial pressures. It is generally recognized that companies engaged in corporate venturing should do it for the long haul – certainly more than five years. Barring this commitment, corporate funding is unlikely to be profitable.

CEO cannot be blamed for trying to keep the ship afloat before investing for the long term, but the temptation is to postpone the long-term investments, which they claim are so crucial. Similarly, governments cut investments in education in times of tight budgets: the saving is felt in the near term, while its impact on the quality of schools and universities will only be noticed much later when the governments will no longer be in power. Parallel to what happens at the company level is the paradox of governments when it comes to technical innovation: numerous ministers and politicians in European countries frequently talk about the importance of investing in technical innovation to boost their countries' competitiveness, but, when money is tight, one of the first budget items they reduce is the investment in R&D. This contradiction is illustrated by the position of several European governments, including France, which for more than 20 years has been announcing its objective of bringing its R&D investments up to 3 per cent of the country's GNP within five years. Empty promises, for in that time both the government and the parliamentary majority are likely to have changed.

It may therefore be argued that the financial pressures in the architecture of the Western system do not provide powerful incentives for the CEO to be a champion of innovation. In the best of circumstances, innovation *may* be at the centre of the radar screen for top management when economic conditions are favourable and the company is healthy and profitable. For this to occur the CEO must be secure at the helm with the personal conviction that innovation is critical for the future of the firm.

The Courage to Champion Innovation

Aventis Pharma Australia claims in its Statement of Corporate Values: 'Courage is a central value for creativity and innovation'. A certain number of CEOs are exceptional in their efforts to keep technological innovation a top priority. They represent a wide range of technology companies in different industries: the examples of Intel, Nokia and Samsung Electronics will be discussed in subsequent chapters. Below are specific examples of CEOs who are passionate about innovation for their firms:

Bombardier: Bombardier is a Montreal-based company with a $8 billion sales volume. Years ago Laurent Beaudoin, now chairman of the company, took two 'bet the company' decisions. He decided to launch his company into mass transit trains, then purchased Canadair in 1986 and developed an aircraft business. On innovation, he says: 'It is my role to always push for more innovation. Whenever I meet the groups at strategic orientation sessions, I ask: what are you coming up with?'[3] Bombardier is now the world's third manufacturer of aircraft, after Airbus and Boeing, at the same level as Brazil's Embraer. The *ethos* of Bombardier has always been to fuel aggressive growth – 20 per cent per year in recent years. Profitable in 2001, it had a Can$615 million loss in 2002, because of the difficult times for the aircraft industry. As a result, Bombardier is now considering selling its historical division of snowmobiles.[4]

Boehringer Mannheim: This used to be a diagnostics company, part of the Corange group. It was bought by Roche in 1997. Dr Gerald Möller, CEO of the company at that time, says: 'I constantly scout for innovative approaches. Sometimes, I detect a small thing, which can make a big difference in the effectiveness of the product. It is therefore critical to remain visionary and open minded'.[5] Consistent with his credo, Gerald Möller is now a principal at the venture capital company HBM BioCapital in Heidelberg.

Canon: Japan's leader in opto-electronics, has a history of disrupting businesses with groundbreaking innovations, such as low cost, personal photocopiers, or the BubbleJet cartridges used in HP-Compaq printers. The company has a policy of aggressive patenting and rewarding employees for new patents. As quoted in Tom Peters' *The Circle of Innovation*, Canon's former CEO, Hajime Mitarai once said: 'We should do something when people say it is crazy. If people say something is good, it means someone else is already doing it'.

General Electric: General Electric's new CEO, Jeffrey Immelt, is re-invigorating the importance of creating revenues through technology. One indication of this is his decision to unleash the intellectual resources of the Niskayuna Corporate Laboratory for the longer term. Many innovation projects now have a two-year horizon. This compares with the previous situation, when researchers were expected to file progress reports every quarter.

Hitachi: Hitachi is one of the most powerful industrial groups in Japan, with an annual turnover of $67 billion. The group is engaged in an ambitious restructuring plan which includes setting up an aggressive programme to commercialize more effectively the enormous scientific and engineering knowledge of the group. As Dr Nakamura, Senior VP Technology says: 'Technical innovation is a central asset for our group. We are leveraging this to build very promising new business, which will bring growth and enhanced profits to Hitachi'.

Medtronic: This $5.5 billion medical technology firm is best known for its pacemakers. It is also active in helping patients with other chronic diseases such as neurological disorders and diabetes. The firm culture highly values engineering and innovation; its former chairman and CEO, William George, states 'It is vital to balance the long and short term growth opportunities. This requires a lot of R&D dollars and a strong managerial discipline to direct a consistent, well balanced R&D programme. It is wrong to focus solely on maximizing shareholder value. If this becomes a key driver, it will force short term considerations to dominate all decisions and squeeze out all intemediate and long term growth opportunities'.[6]

Microsoft: The outstanding success of this star of the 'new economy' owes much to the relentless energy supporting the business vision of its top management, Bill Gates most of all. It also results from his interest in innovation. Each year the company invests $4.3 billion in R&D, which represents more than 15 per cent of its yearly sales. Bill Gates' action speaks louder than words: he stepped down from the CEO position to be more involved with the strategy of the technical development activities of the company.

Saint Gobain: This is one of the oldest companies in the world (founded in 1664) and leads in the production of flat and hollow glass and engineering materials. It purchased Norton in 1990 and revenues were Euro 27 billion in 2002. It has had the same CEO, Jean-Louis Beffa, since 1986. For him, organic internal growth through innovation is a critical priority, which he keeps reinforcing in meetings with his management and staff. 'The CEO must be personally deeply invested in understanding the technology. He must also protect the mavericks', says Jean-Louis Beffa,[7] who is fully engaged in the process of allocating 25 per cent of the R&D budget to long-term projects. These projects are carried out by teams located at the firm's various laboratories. In this way, there is no corporate laboratory as far as facilities go; there is, however, a corporate innovation function. In such developments, which run for several years, synchronizing developments with the needs of the customers is a key challenge. For example, a multi-layer car windshield, with stringent mechanical and optical properties, must be ready when the car-maker is ready to use it in a new model, as part of the branding of the new car.

Swatch: When the inventors of the Swatch, Mock and Mueller, brought the rough sketches of their new concept for a watch to Dr E. Thomke, he immediately saw the potential. Thomke set the demanding target of 5 Swiss Francs (roughly $3.5) production cost for the new product. He then tirelessly supported the inventors in their efforts to develop the revolutionary watch, as well as the automated manufacturing process to produce it. The overall development took two years. Building on this technical breakthrough, and in spite of a disastrous market-test, the marketing and brand-building campaign was effectively executed, to produce today's success of over 400 million watches sold worldwide, as of early 2003.

Valeo: Noel Goutard, former CEO (now chairman) of the automotive supplier Valeo is obsessed with a *will to grow*. During his tenure the sales volume of the company jumped from €1.8 billion in 1987 to €10 billion in 1999. This was primarily achieved by organic growth. Constant innovation for growth and value creation is one of Goutard's five critical basic management themes, together with staff commitment, excellence in manufacturing, total quality and suppliers' integration.

These examples highlight the various ways top managers express their commitment to innovation and put it into practice. More often than not, these top managers have been in charge for a long tenure, ten years or more. This removes the difficulty of maintaining a commitment to innovation programmes, in spite of leadership changes. These CEOs have been able to translate their personal commitment to innovation into convincing their boards about the necessity of investing for the longer term and they have a track record to show for it. They demonstrate that there are exceptions to the 'benign neglect' for technical innovation. There is no innovation paradox in their case: their action directly reflects their conviction.

Innovation in Family-owned and Private Companies

To what extent does the structure of the firm's ownership encourage a longer-term view fostering innovation? How do publicly traded companies that have many shareholders compare in this respect with those that are either private or controlled by a single family? The family-owned public corporations, some of which, such as the German home products company Henkel, are very large, as well as the private – unlisted – companies, are, it is argued, sheltered from 'the tyranny of the short term' imposed by the relentless financial pressures of the markets and the financial analysts. Does this mean they are more likely to take the risk of investing in innovation for the longer term?

If the company's ownership is concentrated in the hands of the top manager, there is at least one consequence: the manager may rapidly take high-risk decisions. An example is the Serono owner-manager, Ernesto Bertarelli, who, within hours, decided to go ahead with a $35 million study to prove that the company's drug Rebif was more efficient than the contender on the US market. The stakes were high, since if the study had not convinced the US FDA – Food and Drug Administration – not only would the US market have remained closed to Rebif, but other markets would also have been compromised.[8] The gamble paid off and Rebif was granted preferred status by the FDA. It is likely that a similar risk would not have been taken by a conventional pharmaceutical company. Certainly it would have taken much longer to reach a decision.

Compared with firms normally listed on the stock exchange, family-controlled companies seem to follow a much less 'stop and go' business strategy, as the corporate memory is better embedded in the management; a down period in the economy is less likely to cause an overreaction, because 'we have survived worse periods in the past'. Because of their longer-term view, such firms are more likely to follow an unconventional strategy. Longer term, in this case, may mean 25 to 30 years, that is, into the next generation.

Family-controlled companies are usually characterized by a high level of trust between employees and the management.[9] This allows intensive debate before making decisions, as well as more alignment in implementing decisions. Again, this climate is favourable for developing groundbreaking innovations to further build the competitiveness of the firm.

Private companies – those not listed on the stock exchange – should not misuse the absence of the healthy discipline imposed by the scrutiny of the markets. More importantly, they may become restricted in their ability to raise capital to finance their growth. In capital-intensive industries, such as papermaking, this may force the firm either to become public, or to be sold to another company. Finally, being listed on the stock exchange also brings notoriety to the firm.

Private companies, however, also enjoy particular advantages. As already mentioned, in addition to protection from the short-term pressures of the shareholders, such companies avoid a number of expenses and constraints imposed on public companies. These include fees for listing on the stock exchange, as well as the energy and time needed to abide by demanding financial reporting rules. Private companies also escape the need to manage the shareholders and the financial community. Such tasks place high demands on the time of top management and require specialized staff: the resulting costs are not negligible for a small or medium-sized company.

In view of all this, and in order to avoid the excesses associated with the 'boom and bust' roller-coaster ride triggered by the 1999–2000 bandwagon 'hype' for so-called technology stocks, will we see an increasing trend towards private or privately-owned companies? Could public companies go through the expensive and extreme exercise of buying back their own shares, in order to become private? Some observers think so.[10]

Technology start-ups constitute a different category of private firms. For them, innovation is absolutely central to the company. They exist to develop it and to bring it to the market. In this case there is no innovation paradox, as all energies are mobilized towards this goal. 'Private equity' investors, which include 'family, fools and friends', as well as business angels, banks, funds and venture capitalists, bet on the energy, good judgement and business flair of the start-up team to build the business. The hoped-for return on their investment is secured on the occasion of the 'exit'. This consists either in selling the firm to another company or investor in a trade sale or in introducing the firm on the stock exchange; this is called an IPO – Initial Public Offering. A successful IPO has a high 'leverage', that is, a high ratio between capital raised and investment. Venture capital funds need such an occasional 'winner' to compensate for the failures of other investments. The period of 2000–2003 of dropping stock markets painfully strained the venture capital industry, since these low levels meant that private equity investments did not have an IPO 'exit'. This meant that introductions to the NASDAQ dropped from $52.6 billion in 2000 down to roughly $3 billion in 2002.

A Swing of the Pendulum?

Making innovation more central to technology companies requires a shift of perspective. It is a tall order to influence the 'invisible hand of the markets' to move the financial pressures of these markets to encourage longer-term perspectives. The shift could come from an evolution of the governance in technology firms. Board members should become more 'innovation activists' and advocate to the share-holders the cause of building the competitiveness of firms through longer-term investments in innovation.

In recent years, financial markets have gone through turbulent times. Trillions of dollars in market capitalization were destroyed between the peak in March 2000 and early 2003. Microsoft's announcement, in July 2003, that it will not distribute stock options any more constitutes a symbolic end to this 'exuberant' era. It can be argued that the global financial system may be reaching its limits and is due for serious reforms in order to avoid the conflicts of interest and excesses of the recent past. Some observers wonder whether, not only managers and auditors, but also bankers, may go to jail. On the other hand, how long will it take for the financial community to forget the excesses of the recent 'bubble'? A swing of the pendulum is far from certain.

The boom years prior to the peak saw the conformist selection of smooth, media-friendly, corporate leaders, largely concentrating on the logic of the firms' financial results, as well as on the ambassadorial duties of the job. They talked about the power of technical innovation, but did not follow through by 'walking the talk' to make it happen. In order to resolve this paradox, boards need to select and support CEOs for their commitment to being closely involved with innovation and longer-term investments; a number of the board members need to be real 'innovation activists'. This pressure should come from shareholders. With the exception of the pharmaceutical sector, however, shareholders rarely ask questions about the innovation policy and the long-term R&D strategy of the firm. A gradual education process on the importance of these issues could improve this awareness.

In addition, some countries are considering forcing institutional investors to vote at shareholders' meetings so that they would take a more active interest in the running of the company and would want to influence it. By and large, such investors only 'vote with their feet' by buying and selling shares in the firms.

North American pension funds are among large institutional investors in multinational companies. One of the largest pension funds is Calpers. It has 1.3 million retired civil servants in the State of California. Any shift in the policy of this fund would have a global influence, since 30 per cent of the $81.5 billion investments (as of 1 January 2003) are outside the USA.[11] If a fund of this importance openly stated support for longer-term innovations, this would influence the companies in which it has invested. In addition, other funds would follow suit; other shareholders would also be influenced and this would contribute to creating a different attitude towards innovations. It can well be envisaged that the Calpers fund would lobby for such changes, as it has already taken stands on less directly business-related, societal issues. For example, on 15 April 2003 the chairman of the board of Calpers wrote to the CEO of GlaxoSmithKline, of which the fund owes one percent of the capital, in order to ask the pharmaceutical company to make every possible effort to make available their Aids treatments in Africa at substantially reduced prices.[12]

Public opinion may also foster a transition. Such an evolution has recently taken place regarding the level of 'golden handshakes' granted to departing top managers. Until early 2001 very generous indemnities to departing CEOs were common practice. AT&T's John Walter, deemed not qualified as CEO, received a $35 million going-away gift in 1998. In contrast, that same year, Mr Kitaoka resigned as President of Mitsubishi Electric, with only his normal pension. In 2002, it was learned that ABB had secretly granted $70 million to former Chairman and CEO Percy Barnevik, more than ten years previously. The troubled supermarket group Ahold gave its ousted CEO Cees van der Hoeven a payoff bonus of €2 million. The list could go on.

Shareholders, including public figures such as Warren Buffet, are voicing concerns over such 'extravagant severance pays'. This is done, in particular, through shareholders' associations. Their general opinion is that CEOs have high pay partly because theirs is a risky job; there is no need for them to have an additional risk bonus. Many recent shareholders' meetings have had tense moments as a result of pointed questioning concerning the remuneration package of top management. For the first time in the UK, shareholders voted against raising the remuneration package of the CEO of the large pharmaceutical company GlaxoSmithKline, on 19 May 2003. Such a vote was extremely narrow, but one may wonder whether this is a precursor of more rebellions in the future. Companies' annual reports are beginning to give information on these items.

Certain CEOs are taking spectacular steps, such as Sidney Laurel, CEO of the US pharmaceutical company Eli Lilly, who took only a one dollar salary and no bonus in 2002: 'If shareholders suffer, then the management must hurt too', he said. In 2003, IBM's Sam Palmisano asked his board to take back his CEO bonus and to distribute it to the 20 executives on his top team.

Such a change of mind about the very specific issue of golden handshakes granted to CEOs took place over a short period of time. A similar swing might take place in favour of the more complex item of longer-term innovation investments. This swing might result from the pressure of society at large, but mainly from stakeholders wanting to create a climate of building true value of firms over time. This could happen if we can learn from the excesses of the 'technology bubble' and from too exclusive a fascination with shareholder value.

In order to reinforce a shift towards a longer-term, innovation-friendly perspective, appropriate rules and regulations are helpful. It is not possible to legislate common sense and innovation any more than integrity, but incentives could be built into the CEOs' compensation policies. For example, instead of connecting compensation to the share price level or the results of the past year, this could depend instead on the firm's results in the next six months, or one to two years in the future. Future results of the firm reflect the impact of today's decisions of the CEO. Another avenue would be not to make the top management's pay increase depend on the firm's improvement in performance over the past year. Instead, the remuneration should depend on how well the firm has performed in comparison to the industry average that same year. This 'benchmarking' could provide a fairer basis for performance evaluation.

Conclusion

In brief, the innovation paradox in technology companies is that, although the innovation process is so crucial to them, the many tasks of the CEOs take them away from fully involving themselves in making this process work effectively. Exceptions in this area act out of conviction; they tend to have a longer tenure than average. They are supported by their firm's traditional commitment to innovation.

How to keep the tyranny of the short term from excessively constraining the breathing space for longer-term innovation? One radical way is for companies to remain, or to become, private; it may well be that the emerging trend in this direction in the USA will continue.

For public companies, resolving the paradox will require nudging the system. This includes aligning the CEOs' incentives with this objective, having more technology-literate 'innovation activists' on the Boards, as well as real support and involvement on the part of influential shareholders. While keeping the healthy contribution of the discipline of the financial markets, the aim is to move away from excessive 'short termism'.

Notes

1. *Neue Zurcher Zeitung*, 27 February 2003.
2. *International Herald Tribune*, 27 December 2002.

 3. *The McKinsey Quarterly*, 1997, number 2, pp. 1–27.
 4. *Le Monde*, 8 April 2003.
 5. Private communication, 4 February 2003.
 6. Medtronic IMD case study 3–1139 (January 2003).
 7. Private communication, 11 February 2003.
 8. Private communication, 18 February 2003.
 9. John Ward, *Families in Business*, April 2002, pp. 74–85.
10. *Ibid.*
11. www.calpers.com
12. *Wall Street Journal*, 22 April 2003.

CHAPTER 3

Is Innovation Manageable?

'Les idées sont pour moi des moyens de transformation – et par conséquent, des parties ou moments de quelque changement.'

(For me, ideas are means of transformation – indeed fractions or moments of some kind of a change process.)

Paul Valéry, *Monsieur Teste*[1]

Some have wondered, not without irony, whether it makes sense to even attempt to manage this wild horse of innovation. Although technical innovation relies on the unpredictable act of creation, management tools can be used to help develop more informed judgements on specific aspects along the idea-to-market process.

As a creative act, innovation is difficult to anticipate. In many ways, the journey of the technical innovator is similar to that of the artist in front of the canvas. Going from idea to market is a complex, intuitive, zigzag process involving many uncertainties that arise from the markets, the existing competitors, possible new players and technology. Uncertainty is at the heart of innovation. It compounds three types of uncertainties: markets, business model and technology. Imagination must read and interpret the reality in order to build a business case for the ongoing innovation project. It is no small task to exercise good judgement when evaluating the risks attached to each of these areas; certainly these can be reduced if the project benefits from a high level of commitment and leadership. If this is the case, the chance of success could well become a 'self-fulfilling prophecy'.

As seen in the previous chapter, the CEO has a strong 'top–down' role in putting innovation at the centre of the village, but there is also a need for powerful 'bottom–up' dynamics in proposing ideas and undertaking the idea-to-market journey. In recent years, a number of tools and management practices have been increasingly used to enable the course of the innovation journey. These tools focus on innovation developments carried out within the firm. They help foster and stucture exchanges of ideas, data and judgements in order to enable more informed decisions regarding innovation projects.

The Act of Creation

Creativity is celebrated by all societies and cultures. It draws on inspiration, intuition and passion to make an impact on the world. It is at the centre of the work of artists and scientists alike. For them, the 'self-starting' quality of their creative energy is very similar to that of entrepreneurs.

In his book *The Act of Creation*, Arthur Koestler describes the apparent sleepwalking approach (in fact, the French title of the book is *Les Somnambules*) followed by great discoverers such as Johannes Kepler in their quest to understand the laws of the Universe.[2] As Kepler was developing the equation describing the elliptic trajectory of the earth, he apparently made two mistakes which cancelled each other out, so that he arrived at the correct mathematical formula as if guided by a mysterious intuition.

A similar intuition guides scientists and artists alike. In their efforts to 'blaze a trail', as they seek to have an impact on the world, such individuals focus on their desire to fulfil their project, whether it is a piece of research, a piece of music, a part in a play or a work of art. They all have the same 'self-starting' quality: they are passionate about their act of creation. They all have the same need to concentrate on their creative work and to draw on their inner strengths to enable their activities. In many ways, the artist embodies the model of an entrepreneurial outlook in his or her efforts to shape the outcome by marshalling emotions and skills. As Francis Bacon said in *Advancement of Learning* in 1605: 'He that will not apply new remedies must expect new evils; for time is the greatest innovator.'

The gratitude of society towards creators and innovators is not very strong, if measured by monetary reward. With the exception of a small number of 'stars', artists tend to have low revenues and precarious positions; our societies do not generously reward the creativity of those who most inspire us. There is a similar 'pecking order' for income within the firm: at a comparable hierarchical level, managers in science and technology are paid less than financial officers. True, the motivation of technical knowledge workers largely comes from being involved in a challenging project. For them, however, the salary level is a very close second motivating factor.

In order to reward the innovators who contribute to the business success of their employer, large technology companies are increasingly giving bonuses. The days have gone when the author of a patent was grateful to receive a symbolic dollar bill or an engraved watch on the occasion of an inventors' ceremony when the patent was granted. Intel gives a bonus in the order of $50 000 to authors of an invention significant for the business of the company. Since it is impossible to know in advance the value of a patent, it is best to reward the inventor later, when the benefit of hindsight will reveal the extent to which the patent has contributed to the firm's business growth. The inventor is then rewarded in relation to this contribution.

In Japan, many technology companies are rewarding inventors who come up with value-adding ideas. Hitachi, for example, is considering a new bonus system to reward its inventors, based on the impact of the patent on the competitive position of the firm in the relevant market segment. A decision by Japan's Supreme Court on 22 April 2003 is likely to renew the debate as to how high a reward should go to employees who are authors of commercially important patents. This decision required Olympus to grant a much better reward (an additional $19 000) to the inventor of an optical device, who claimed that the $1700 he had initially received from his employer was insufficient. For many years, Japan and Germany have had laws requiring that firms appropriately reward employees who are authors of patents. What is meant by 'appropriate' is a matter of judgement and is in the process of being re-evaluated.

Uncertainty is at the Heart of Innovation

Each innovation project is unique. It represents a gamble; it is an option on the future. Nobody knows for sure how it will turn out; if its outcome were known, there would be no need to carry it out. Innovation projects have widely different scopes. They might concern the development of a new Airbus airplane, which would involve innumerable sub-developments. Another innovation project might involve the development of new software or an improved catalyst for the chemical industry. Uncertainty is at the heart of technological innovation. From this flow several characteristics of the culture in the R&D function.

First, there is a tendency for staff to compensate for the uncertainty inherent in the nature of their work by demanding they have stability in the organization and work setting. This is apparent in periods of transition: the innovation pipeline dries up during periods of organizational change, for example, when one company is purchased by another or if there is a merger, such as between Glaxo and Wellcome in 1995. All efforts were made to implement the merger rapidly and early in the process the number of technical professionals was, as usual, 'streamlined'. In this case, the headcount of the combined R&D function went from 13 000 to 10 000.

Similarly, when Pharmacia was acquired by the $57 billion drugmaker Pfizer, it was announced in April 2003 that three out of the twenty-five research centres of the combined company would be closed, affecting several thousand employees. Mergers offer an opportunity not only to eliminate duplication of efforts, but also to refocus the strategy, including a readjustment of the thrusts of the R&D activities to prepare better for the future. In such periods of transition, employees in general, and R&D professionals in particular, put their creativity on 'standby' and their output drops dramatically. The question for management is then how to jumpstart the innovation heart so that the flow of innovations starts to circulate again as quickly as possible. A large part of the answer lies in management rapidly clarifying the organizational structure, the new mission and strategy of the merged firm. This must be credibly and fully communicated to all employees. A great deal of time must be dedicated to dialogue in order to convince the staff. When the message from the conductor has been well received, as it were, the musicians begin to play, and the motivation and productivity of knowledge workers begin to rise again.

Second, R&D professionals overcome the uncertainty of their jobs by drawing on the energy fuelled by their own quest. In curiosity-driven, as well as in business-motivated R&D, they are obsessed by the urge to shape the world around them, whether it is the performance of a software or a new treatment against a disease. This positive stress, as well as stretched goals, stimulate researchers into becoming 'finders'. It was by responding to the king's pressing demand to devise a way to make sure that his crown was made of high-carat gold that Archimedes discovered the buoyancy principle. For each triumphant 'Eureka', however, there is much frustrating hard work. In addition, management sometimes seems insatiable; when, after painstaking efforts, an alchemist discovered the formula for Meissen porcelain, which brought enormous wealth to Saxony, the king said: 'Very well done! Now, how about making gold?'

Let us assume that, inspired by the CEO and encouraged by the company culture, the company is blessed with a drive for innovation. A well-known model in this area is the Minneapolis-based company 3M. In order to foster an innovation culture, 3M instigated a much-discussed measure, namely the 15 per cent rule. This provided R&D employees with a motivation for experimentation, by allowing them to use this fraction of their working time to pursue their own projects. The company jargon talks about 'bootlegging', an expression which conveys notions of underground, entrepreneurial, tenacious activities. Indeed, some of these 'pet projects' led to nice business-making new products, such as the celebrated post it notes.

Clearly, one single measure is not enough to help promote an innovation culture. A set of consistent measures must be taken. What is remarkable about the innovation-friendly 3M company is precisely that the many different measures taken are consistent and reinforce the other, while management is aligned with them. Then, and only then, does the possibility arise of creating and maintaining a strong innovation culture.

Having this positive culture is necessary, but still not sufficient. Together with high staff motivation, a risk-tolerant culture constitutes only a terrain on which innovation might emerge and flourish. But how effectively it will develop into commercially successful products is a question that has concerned managers and business schools, particularly after they noted the remarkable successes of a handful of Japanese technology companies in the 1980s. How, they wondered, could they improve the success rate of bringing technical ideas to market? Japan's steel-making and automotive industries started pointing the direction, with companies such as Nippon Steel, Honda and Toyota, as did microchip and consumer electronics companies such as Canon, Matsushita, NEC and Sony. These are the well known names which, by and large, continue to excel in bringing successful innovations to market in rapid succession. Many other Japanese firms are also very successful at combining excellent abilities in the technical and business spheres. Some are active in the life-sciences area, others in robotics and 'mechatronics', which involves combining precision mechanics and electronics in products such as knitting machines and small motors. These companies are not so well known; examples include Hayashibara, Noritsu Koki, Shima Seiki, and so forth. The sluggish economy in Japan during the last decade has led the West to assume that nothing good is coming out of that country. This is a great mistake; in Japan, many manufacturing companies are very astute in exploiting a market niche by combining technical excellence and good business sense. Moreover, some of them are cleverly exploiting the advantage of keeping technology developments in Japan, while locating their production facilities in nearby China.

Before talking about the practices applied along various phases of the innovation process, it is useful to have an overall visual concept describing this process. The metaphor most often used is that of a 'funnel', as shown in Figure 3.1. The steps go from idea to feasibility studies, pilot-scale testing, scale-up, production, ramp-up and market launch. Encircling the funnel is a spiral, suggesting the constant iteration of the innovation between market and technology throughout the process. The early part of this trajectory is about evaluating ideas, enriching them through discussion and finding additional information for assessing them. It is an open, exploratory phase. In contrast, managing a project that is nearing the market launch may be characterized as being carried out with *disciplined speed*. The rules of rigorous project management are critical at this stage: in other words, the management style at the exploratory research, as in the case of the search for a NCE (New Chemical Entity) in the pharmaceutical industry, is very different from the 'downstream' management of clinical studies where disciplined project management is key. The molecule-to-drug development follows an innovation process that is codified by regulatory institutions such as the US FDA – Food and Drug Administration, or the European Agency for the Evaluation of Medicinal Products (EMEA), its European counterpart based in London. This process is illustrated in Figure 3.1. It begins with research aimed at identifying a molecule active to treat a condition. This phase requires massive capital investments in genomics, proteomics modelling, combinatorial

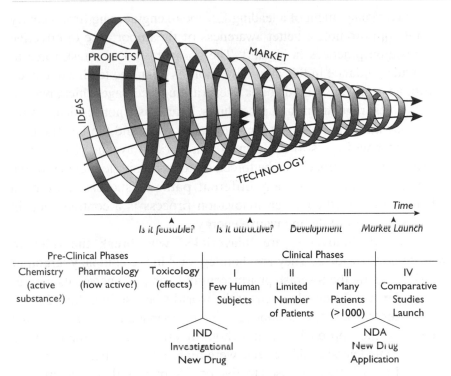

Figure 3.1 The innovation funnel, with its various steps. Also shown is the example of the regulated drug development in the pharmaceutical industry.

chemistry and high throughput screening of the molecules. It continues with checking the efficacy and the absence of negative side effects. The overall process costs close to $800 million per NCE and lasts between ten and twelve years.[3]

Along the idea-to-market trajectory, decision points provide an opportunity to evaluate, modify or screen. It is highly desirable to have elements for enlightened selection of the various innovation projects as early as possible: in this sense, '*early failures*' are wanted. It is important to be in a position to discontinue projects as soon as possible, in order to free resources for potentially more promising alternatives. In this constant effort to optimize the allocation of resources, a company having an innovation machine with a slightly better success rate than its competitors' holds a powerful advantage. The cost of continuing to invest in a dead-end project is high; a more promising project may be put on the back burner as a result and precious time thus wasted in the competitive race.

The management of a leading European engineering firm recently wanted to promote a better awareness of the importance of effective innovation practices in its staff. The company asked a task force to provide guidance in this area. Five months later, the team came back with an 'innovation guide book', several hundred pages thick, packed with rules, regulations and procedures. This was a nonsensical result, as innovation progresses in unpredictable ways and the last thing an innovator will do is to conform to a prescribed avenue. There is no way a single approach can be forced on to every single member of the staff, since there are many different paths to success. Forcing a bureaucratic method on an innovation process is a contradiction in terms; it will kill the innovative energy.

Indeed, innovators are 'mavericks' who break the rules to progress their ideas. They are 'bootleggers' in the company jargon of 3M. At Astra, the Swedish pharmaceutical company before its merger with Zeneca, the development of the world's best selling drug, Losec, for ulcer treatment, was stopped twice by management; the first time because the approach was not thought effective; the second time because undesirable side-effects were feared. Each time, the inventor, Sven-Erik Sjöstrand, refused to give up. He bypassed or overcame the objections and pushed the project forward, obtaining along the way a Swedish government grant, which saved the project at that particular juncture. In this case, breakthrough innovation was the result of both a nonconformist culture encouraging an open dialogue with external specialists, and the intuition to follow a different therapeutic approach and a tenacious 'can do' attitude. It is impossible to legislate an innovative culture. It takes a long time to build and must be continually strengthened; it can, however, be destroyed very quickly.

Whereas the fertile terrain of an innovative culture has to be nurtured, technology development tools and practices can help the innovation process along. They help enable and guide this process. These tools have been increasingly used by companies since the mid-1980s, when the phrase 'technology management' was coined to designate the specific challenges of turning technology into competitiveness, wealth and jobs. The rationale for companies was to improve their innovation process, as compared with competitors. They also aimed at shortening the duration of the time to market. For example, in the late 1980s most pharmaceutical companies launched aggressive initiatives to reduce the drug development time, with the objective of commanding higher prices in the early years and increasing the time during which the patent protects the drug in the market before it becomes a generic molecule.

Some of these tools for managing innovation are listed below; (a more detailed description of these tools may be found elsewhere):[4]

- Multi-functional projects, project management and 'commando' projects

- Innovation board/council or technology management cell

- Risk-difficulty/return analysis and portfolio of projects

- Competitors' analysis and environment tracking

- S-curves

- Technology-business mapping

- Focus groups, quality function deployment

Multi-functional Projects

Managing an innovation project in a truly multi-functional way constitutes the most powerful way to increase both the speed and the effectiveness of bringing innovations to market without gaps. Such a smooth, integrated process is sometimes called 'seamless'. In such projects, staff from R&D, marketing, and manufacturing, drawing on expertise in design, financial management and patents, work together in an integrated and organic way. This diversity of skills and experiences is placed under the leadership of a project manager, who will progress the innovation to the market.

An example of such project organization is the Dutch photocopier maker Océ which for the two to three years required for the development of a new photocopying machine assemble a team of 40 to 50 members who report to a project leader. In order to intensify communication among the team members, they are asked to move to a common location in a large open-plan room. As is well known, proximity of team members creates a much higher probability of formal and informal contacts, in contrast to situations where team members work in different rooms, even if the rooms are relatively close to each other. The visual contact provided by an open plan tremendously enhances communications among project members.

At the end of the project, team members go back to their original departments. This 're-entry' generates stress, since members wonder whether their next project will be challenging. Allocation of personnel will depend on a three-way political conflict: the staff member in question, the project leader and the department manager, who may well have different views on the matter. As in 'matrix' organizations, there are very often tensions between project and department managers. In strong project organizations, management usually favours the project leader.

Even in today's world, where we increasingly manage in electronic space, geographical proximity remains a powerful tool to foster communication. This is particularly the case in the course of developing a complex product such as a new car model, with its thousands of parts. In order to group together the various actors in the development of a new car model, automobile companies have built large 'technical centres'. These include the Munich-based BMW development center, as well equivalents at DaimlerChrysler, Ford, Peugeot and Volkswagen. The basic objective for such a center is to enhance the ease and the richness of face-to-face communication among the many persons concerned with a single project, who were previously scattered in different locations. The quality of output, the speed of the development of new car models and the cost saving are expected to provide handsome returns on such large investments in technocenters.

Milestones, as well as criteria for stopping or continuing innovation projects, must be clearly defined, consistent with the business objectives of the firm. The paramount element for the success of a project is indeed the quality of the project leadership. Good project leaders master the people issues as well as the technical and business aspects. They have to be effective leaders of people, able to manage across functions and cultures, as well as location and geography; team members, often from different partner-companies, are sometimes located in different parts of the world. Communication technologies constitute a particularly powerful enabling tool in this case; the effective project leader must also be a good manager in 'electronic space', that is, using these technologies as assets to manage the project, rather than as a mere channel for data transactions. Numerous software products, as well as hardware equipment, now exist to help project management. Electronic space is particularly well suited in the case of software development, for which several teams in various parts of the world may collaborate on the software code. The team in Tokyo thus hands the day's work to another 'shift' in Paris, who sends it on to Palo Alto at the end of their day. Leveraging time zones in this way, in effect, allows a round-the-clock development, sometimes called 'the 24 hour laboratory'.[5]

The multi-skill profile required of a project leader represents one of the major bottlenecks for technology companies. This scarcity of effective project leaders is often critical when such managers are entrusted with 'bet the company' projects without either proper support from management or sufficient training. Managers of such development projects are very much like 'mini CEOs', in the sense that they truly have a general management role and their success or failure means a great deal to their firm. The management development of project managers must have a high priority for technology companies, as will be discussed in Chapter 7. It represents a long process, in which coaching, job rotation across functions, as practised in Japan, as well as wide diversity of experience are all contributing factors.

Most companies have adopted the principle of multi-functional teams. Going from this principle to the practice of truly organically integrated teams is a challenge that has met varying success. Examples include the development of a new drug, such as Losec, as mentioned earlier, in the pharmaceutical company Astra, a new mobile phone by Samsung Electronics, or a new car model at Renault, where the *Twingo* was developed with a purposeful and disciplined approach to project management.

An extreme version of this approach is the *commando* project, sometimes also called skunkworks. This is used when a company must make a major effort to catch up with competition in an area critical to the future of the company. For example, in the late 1970s IBM missed out on the strategically crucial product of the personal computer to a brilliant pioneer, Apple, with its Macintosh. In order to make up for lost time, IBM sent a small team, composed of a dozen high performing staff, to Florida, far from the distractions and bureaucratic grind of the Armonk headquarters in New York State. With full support from top management, the team was given the task of rapidly developing a PC prototype that would put IBM on the map for this type of product. The project was successful and IBM started selling PCs in the early 1980s.

Many other companies have used this temporary commando 'strategic catch-up' exercise, which typically lasts between 18 and 24 months. They include Sun microsystems and Sony for developing workstation computers. At Hitachi high priority projects that have strong support from senior management are called *Tokken* projects. In order to increase the choice of options, management may elect to have two commando teams run in parallel and compete. This duplication of efforts is expensive, but may be justified as an insurance to get the best possible quality of outcome. It was purposefully practised in the case of Korea's Samsung Electronics, discussed in Chapter 5, at the early stage of their 'bet or break' development of DRAM chips. In this case, parallel teams were set up in Seoul and Palo Alto. Both teams were given the same objective of developing a new chip. They were told that they were competing for the same task. On several occasions, the Seoul team came up with the solution that was finally retained.[6]

The Innovation Board

Technology companies have a portfolio of innovation projects underway, but where does responsibility lie for guiding and prioritizing these projects? Personal politics? Chance? Who plays the role of sorting station or 'control tower' to decide which project should take off before another one for the good of the firm? In order to oversee the ensemble of the most critical projects, companies have put in place innovation boards. These are composed of a small group of five or six people who combine the necessary technical and market knowledge. These individuals typically tend to be vice president of technology, senior R&D manager, business unit manager, business development executive, marketing managers, and so on.

With a keen eye on the external competitive and business environment, the innovation board helps supervise, guide and prioritize the progress of innovation projects most critical for the future of the firm, those which are likely to bring maximum returns, as well as platforms for subsequent developments. The innovation board typically meets once every few weeks in order to review, critique and reorient the projects in question. It is my experience that roughly 60 per cent of large technology companies have put in place such councils. They function more or less effectively, depending in large part on the level of commitment of the company's CEO to innovation. As always when a new management practice is introduced, the expectations are high and the board members are very diligent. After some time, however, they stop attending innovation board meetings so diligently and go to meetings with clients instead.

In some companies, such a board is really not necessary because their executive committee itself performs this task: a very substantial part of their discussions (well over 50 per cent) are precisely concerned with new product development projects. One such company is Nokia Mobile Phones. In this fast-paced, highly competitive industry, a high tempo of innovations must be followed by disciplined, timely developments and market launch with attractive design and high quality. It therefore makes sense that this company would dedicate much attention to product strategy and development, although most competitors delegate this work to the innovation board.

Project Portfolio Management

Portfolio charts constitute another tool to help set priorities for innovation projects. There are a number of approaches for doing so. One is to analyze the potential benefits versus the size of investment and risk.[7] This allows the firm to regularly rank the development projects under way in terms of potential commercial success.

One thing is clear: based on my experience of working with many technology companies, there are too many projects – between 20 and 30 per cent too many – going on at any given time that do not bring much value to the firm at all. For these projects the returns on investments will clearly be far too small. They may be the result of politics in the firm: the 'pet' project of a board member, for example. Such 'project waste' may also result from keeping technical specialists busy without enough consideration of relevance to the business. This 'knowledge inertia' is very compelling when the firm wants to make sure that its leading scientists in an area continue to be funded for the projects they request.

Streamlining the portfolio of innovation projects should be done continuously and relentlessly. Presumably the innovation board would help in this task. Focusing development activities improve efficiency in several ways: it directs resources down the most fruitful avenues and, by so doing, saves precious time in the competitive race.

The S-curves

An S-curve is a graph which illustrates the correlation between the improvement in performance of a product and the cumulative efforts invested in its development, as shown in Figure 3.2. A new technology – going from analog to digital in mobile phones, for example – corresponds to a new S-curve. Another well known curve example concerns semi-conductors: Moore's law predicts that technical developments cause the circuit density, and its functional performance, to double every 18 months. This empirical rule, based on an observation by Gordon Moore from Intel in the 1960s, is still valid today.

The S-curves represent a useful tool for market and technology staff to exchange their views and perceptions as to the stage of development of their company, particularly as compared to what they know of the activities of their competitors. It allows them to discuss whether their development efforts are going in the right direction and whether the rate is appropriate. Prior to this, the members of the group must define what parameters should be the criteria for their evaluation on the vertical axis: in the case of a television set, for example, is it the brillance and sharpness of the picture or is it the design of the set? In the case of a very complex product, such as a car, are we talking about the design of the vehicle or its fuel consumption? How about the

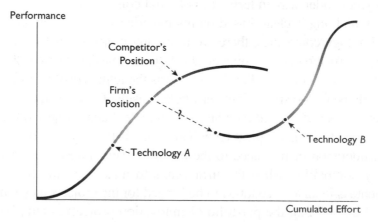

Figure 3.2 S-curves for two different technologies, allowing a discussion of the position of the firm, as compared to competitors. Also illustrated is the difficulty of properly timing investing in an emerging, competitive technology.

case of a software? The selection of the specific evaluation of the 'performance' most relevant to commercial success will produce long and animated discussions before any consensus can be reached.

Switching from one technology to another can also be expressed by S-curves. It amounts to going from one curve, relating to one technology, to another, relating to the successor technology: going from vacuum tubes to transistors, for example, or from film-based to digital photography. The consumer electronics company Sony is a leader in digital cameras; in fact, it was never involved in the previous technology of paper-based photography.

It is extremely difficult for a company to switch successfully from one technology to another at the appropriate time. A newcomer, without the legacy of the past, is often the new challenger, using the new technology. One way to manage a radical technology jump is for a firm to make an alliance with a company that has the emerging technology. The legacy of the past and the accumulated know-how represent considerable barriers to change. Possibly, however, success is the biggest barrier to change. There are exceptions to this observation: the Osaka-based Takeda company is today one of Japan's leading pharmaceutical companies. In the nineteenth century its business was trade in Chinese herbal medicines. This firm had the time gradually to evolve, but it is probable that the present-day fast paced situation would not allow such a transition.

Another example would be Switzerland's refusal in the 1960s for the watch industry to go digital. This decision was mainly motivated by the enormous accumulated know-how in micro-mechanics for analog watches on which the economic activity of the regions of Neuchâtel and the Jura mountains fully depended. To switch would have made this know-how obsolete. This situation led industry leaders to convince themselves that the industry should not go that way and as a result they explicitly dismissed digital watches as 'gimmicks' without a future.

Technology Mapping

A company deploys its technical know-how in as many product applications as possible. The tool of technology mapping shows what technologies are embodied in the products of the firm and helps make the

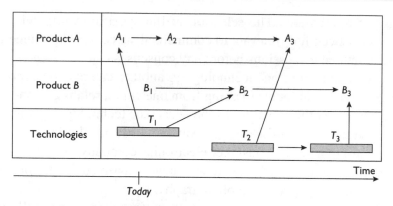

Figure 3.3 Technology mapping connects technologies and products.

connection between products, or families of products, and the technical expertise required to develop them.[8] This is done over a time horizon of three to five years. These relationships are illustrated in diagrams, such the one schematized in Figure 3.3. Products A and B make use of technologies T. The diagram also indicates at what point a technology overtakes another one, affecting specific components or products as a result.

This planning tool thus helps organize the deployment of company technology and maps the best ways to apply technical skills to support the development of as wide a range of products as possible. This notion of 'technology platform' is often mentioned in relation to automobiles: the desire to use the same parts as often as possible for different car models without compromising the different characteristics and branding position of the each car model in the marketplace. Achieving economies of scale through product platforms is used in many other sectors, including consumer electronics items such as DVD players or Walkman, as well as for white goods and personal care products.

Finally, technology mapping helps guide hiring. It is used by firms to identify the technical expertise needed, so that appropriate new talent is hired when it is needed, in order to have time to integrate the new employees into the company and to be ready to carry out the product developments planned.

Quality Function Deployment

The term Quality Function Deployment refers to the aim to relate the demands of customers to the technical features best suited to satisfy them. For many years it has been used in the car industry, where the complexity of the product needed a tool to correlate between desirable features and technical characteristics of the product.

Market studies and customer focus groups help define what features and characteristics are desired in a new car model. These are then translated by the car maker into the technologies that will best satisfy customers and be at the right cost. The car maker identifies the preferred solution, discusses it with its suppliers, with whom a solution is agreed in effect pushing the burden of innovation to the supplying firm. For example, the characteristics and functions for a new design of car lights, or an on-board multiplexing network, benefit from the inputs of customers, finalized by the car maker, which then forwards them to suppliers, from where they will be sourced. Similarly, concerning the general experience of car users, focus groups are asked to give their comments on more diffuse issues, such as desirable smells and noises inside the vehicle. This input is then translated in terms of choice of materials and engineering design.

Innovate with a High Market Orientation

The use of tools such as those described above has greatly helped technology companies make their innovation process more relevant to value-creating in today's world. In particular, these tools have contributed to making the dialogue between technical and business personnel more systematic and fruitful. The former have developed much more of a business sense as a result and are in a much better position to apply their technical expertise in the marketplace.

The tools and practices described help firms conduct their idea-to-market process as productively as possible. They concern the management of the process with its successive milestones, including the decision to continue, modify or interrupt specific ventures. One variation on this theme is the so called 'gate process'. Whatever the name, the crucial point in these milestones/screens is to think very hard about what specific selection criteria must be applied at the various stages of the product development. These include functions, cost, market size, required distribution channels and relevance to the product and brand strategy of the firm. In the portfolio of innovation projects, what is sought is to pick the winners, or conversely, to have *early failures*. Each project is continuously evaluated with the best possible data on which to base the decision to continue or stop the project, and at the earliest possible time. As already indicated, a dead-end project that is allowed to continue represents a high opportunity cost; it consumes resources and time which will not be available for a more promising project.

These management tools are today used in most technology companies worldwide. They do not constitute a panacea in any way, but they do help mobilize the experience and knowledge available in the various parts of the firm. They catalyze the dialogue between the technical and business functions, offering a forum in which to organize and synthesize the information and judgement on a specific business issue in a systematic way. The tools do not make the decisions, but can help make better decisions.

Each of the functions in the firm must cultivate its own specific strengths: R&D cultivates its technical leading edge, scans technology developments outside the firm, while remaining keenly aware of the business and customer evolutions. Each function must remain open-minded and eager to engage in constructive dialogue with other functions, in the best interest of the future success of the company. The worst-case scenario will be if the different functions lose their edge of excellence and are all reduced to a tepid common denominator. Diversity of honestly argued points of view constitute a critical asset. The objective is not to have the marketing function become knowledgeable about technical matters. Rather, the idea is for marketing to be 'technology-literate' enough to be able to ask the right questions of the technicians and to understand them well enough to have a productive give and take. The ultimate aim is to nurture dialogue to make decisions with the best possible judgement. In this sense, the notions of 'market pull' and 'technology push' have lost much of their meaning; the reality is more complicated in that technical aspects and the perceived market reality interact early in the discussions on new product developments. This interaction goes on all along the development-funnel of Figure 3.1, in which market and technology are made continuously to bring their input along the iterative spiral between technology and market.

As a key participant in the innovation process in technology companies, the R&D function has thus greatly evolved in recent years. First, it has made enormous progress towards being much more *market oriented* and having a sharper *business sense*. This means that the technical staff are curious about market conditions and ask themselves constantly about the relevance of their work to the business of the firm and to its customers. In many Japanese technology companies, such as Sony and Canon, engineers and scientists are truly obsessed with understanding the state of mind of customers. This is true even if the planned development takes several years until market launch. In Saitama, one of Hitachi's advanced laboratories, an engineer working on an application of superconductive devices is preoccupied with an in-depth understanding of the evolving wishes of the customers, although this development will not be commercialized for more than ten years from now.

Further, R&D personnel are much more aware of the business implications of technical choices. This means they are motivated to influence the strategy of the firm by interpreting for the business what technical developments imply in terms of opportunities and threats. They have made much progress in making their voice heard by learning and practising the 'businesspeak'. This evolution over the years towards a remarkable increase in reciprocal interaction between business and technology is schematically illustrated in the ternary diagram in Figure 3.4.

In this diagram there are roughly three periods. In the aftermath of World War II, managers of the R&D function were essentially asked to have very high technical competence; in the diagram, the corresponding point was close to the top corner of the triangle. The

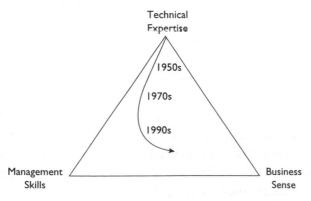

Figure 3.4 Shift in the emphasis of three main elements required for managing technical knowledge professionals.

assumption was that 'good scientists and engineers will eventually come up with products which will make us profitable'. Consistent with this view, R&D buildings looked like a campus, in order to seduce young graduates with good working conditions. Geographically and in mindset, such laboratories were far from the customers. At that time, the idea was to leave the technical professionals happily working largely on curiosity-driven projects in what is sometimes called an 'ivory tower'. In terms of physical location, the Bell Laboratories in Murray Hill, New Jersey, were the prototype of this 'laboratory in the woods'. This place, however, was remarkly productive. It is best known for the invention of the transistor in December 1947. This led to a Nobel prize in 1948 and was the origin of the present massive electronics and software industry.

Later, it was considered that R&D management should include both personal qualifications and administrative handling, so the graph point shifted towards management skills. Laboratories thus experimented with dual management, including a director, a scientist or engineer, former 'master of the craft', assisted by an administrative director. Hospitals had a similar evolution, with a medical doctor and an administrator sharing the overall management. Professionals could thus identify with the medical director, with whom they could discuss contents issues as a colleague, while 'administrative' processes were handled by the administrative director.

Since the mid-1980s the emphasis has been increasingly on market orientation: technical staff must work closely with the rest of the firm on innovation projects, demonstrating a keen business sense and a high awareness of the wishes and dynamics of the marketplace, the consumers and the competitors. Accordingly, the trajectory in Figure 3.4 shifts to the right.

In the recent past, technology companies have followed this trajectory towards a considerably sharper business sense. They now manage innovation projects with a much more market-oriented perspective. It is fair to say that within these companies, R&D is probably the function that has evolved the most over the last 15 years or so. This transition has been caused by enhanced competition in an increasingly interdependent world, the rapid rate of change, as well as structural factors, such as deregulation and privatization, particularly in the area of telecommunications.

Conclusion

In brief, in the uncertain field of innovation management a number of lessons are clear that can be articulated around the metaphor of the innovation process as a chemical reaction. As is well known in chemistry, a reaction takes place more strongly and rapidly as the temperature is raised. The innovation-orientation and the management style of the firm are the equivalent of the temperature that allows the innovation reaction to be initiated and take place. Innovation is fostered by the high temperatures of a rumbustious, non-bureaucratic environment, with a highly committed management.

In the alchemy of innovation, the reagents are the creativity, talent, knowledge and motivation of the employees. The quality of output of the innovation process heavily depends upon the crucial human factor, as will be discussed in Chapter 7. The catalyst enabling the innovation reaction is constituted by the practices and tools just described. These elements enable sharing knowledge and structure a fruitful dialogue between the business and the technology part of the firm. They are helpful in forming a better judgement on all aspects of the innovation process.

These tools have helped the R&D function acquire much more of a business sense, as well as streamlining the firm's portfolio of innovation projects. One of the objectives is to identify failures early. As you use these tools, remember that they are not a panacea. They do not make the decisions for you. At best, they help enhance the quality of the collective judgement of the firm. As a result, the firm will hold a more effective portfolio of innovation projects.

Notes

1. Paul Valéry, Œuvres complètes, vol. 2, p. 71 (Gallimard-La Pléiade, Paris, 1960).
2. Arthur Koestler, The Act of Creation (Arkana, London, 1989).
3. www.csdd.tufts.edu
4. G. Gaynor (ed.), Handbook of Technology Management (McGraw-Hill, New York, 1996).
5. Georges Haour, 'Managing Innovation in the 24 hour laboratory', in Mastering Global Business (FT Pitman Publishing, London, 1999).
6. Georges Haour, Samsung Case Studies (IMD, 2000).
7. G. Gaynor, op. cit.
8. P. Groenveld, 'Roadmapping integrating business and technology', in Research Technology Management, 1997.

CHAPTER 4

Leveraging Technical Innovation through a Diversity of Channels

Part of the evolution described in the previous chapter is that innovation projects increasingly draw on external participants. It is often said that there is no way any given firm can be good at everything and there is much more going on outside the firm than within it. Tapping into external knowledge is therefore a must. The first part of the statement is nothing new; the second part has now become more necessary. Companies, however, are not going far enough in the direction of practising 'distributed innovation', which will be discussed in Chapters 5 and 6.

During the past few decades, the amount of available technology has increased massively, mainly as a result of the considerable activities in science and technology at universities, companies and public laboratories. These sources have also become much more accessible to companies as a result of inexpensive means of transportation and communication. It is quite feasible to have a collaborative development involving, say, a firm in Singapore and a laboratory in Germany; affordable telecommunications and air travel make this possible. In the quest for improving their returns on investments in technical innovation developments, firms must align their practises with this new reality.

This *aggiornamento* means that a firm must mobilize if it wants to generate new revenues from commercializing its technical resources. Agile usage of many different channels for creating value out of technology is illustrated by the remarkable example of Generics, a company fully dedicated to *leveraging technical innovation in multiple ways* as described below. It offers technology companies a model for how better to leverage their technical expertise for growth and profit.

Multiple Leveraging of Technical Innovation: the Example of Generics

Generics is a company entirely focused on technological innovation. Located in Cambridge, Great Britain, it employs more than 150 professionals with advanced degrees in science and engineering. The overall staff is 230 and its extensive facilities include up-to-date laboratory facilities. Generics became a public company in 2000 when it raised £38 million, net after costs and a loan repayment and it is listed on the London Stock Exchange. The company was founded in 1986 by Professor Gordon Edge with the aim of harnessing technical innovation to create value through a variety of avenues for commercialization. The name of Generics suggests the company's practice of drawing on many different generic technologies to power innovation. Its unique business model includes several different interdependent activities.[1]

The first type of activity is to provide *technical services* to companies. This activity makes a contribution to firms by carrying out innovation projects in a number of areas: new product development and engineering-based innovations, but also advising client companies on strategy, business innovation and due diligence prior to acquisitions. All of these activities aim at guiding companies in their efforts to grow from innovation and to shape their own future.

Innovation projects are engineering-based and carry intellectual property rights. The general rationale of these projects is to improve the *competitiveness* of the clients by developing for them new or improved products or processes. On average, Generics files more than 450 patents per year. The laboratory specializes in robust, patented, usually basic science-based innovations, providing potential advantages in the sensor, medical instruments, healthcare and telecommunications sectors. The innovation projects are performed under contract with clients, mainly large companies such as Nike or Siemens. Once a problem has been stated, it is discussed and defined together with the client company. The approach on how to solve it is agreed upon and described in a proposal, which, together with the contract, defines the work plan as well as the intellectual property rights. The contract includes an important section on intellectual property rights: the rights to be transferred to the client are clearly specified, in terms of scope and geography. Generics, as is common practice in CROs – contract research organizations – thus retains its rights to work with other clients outside the specified scope. In this way a common technological platform and expertise may be deployed, in parallel, for developments of other applications or different geographical areas.

The project begins soon after the contract is signed. It is implemented on a best-effort basis; its overall objective is to develop an innovative approach towards a new product or process. The projects are carried out in a multi-disciplinary perspective, in order to provide the broadest toolkit for finding the best solution to the problem. True to the tradition of the University of Cambridge, physics and materials sciences are the central scientific disciplines in the laboratory. Projects deal with developments in topics such as sensors or telecommunication technologies, or physics-related innovations to be used in the medical and life-sciences sectors. As an example, Generics manages a field testing of 3G mobile telephony.

Occasionally, projects are carried out on behalf of public organizations, such as recent work for the United Kingdom's Government Foresight exercise that aimed to identifying the key technologies for the future. Another project resulted from Generics being invited to participate in a European Union project, the objective of which was to provide input on policies that will develop the best practices for incubating technology start-up companies that are coming either from private organizations or from university and government laboratories.

Such technical services and advisory activities provide Generics with a strong market presence. It ensures high visibility to companies throughout Europe and the USA. Its activity is managed to be profitable in its own right. In addition, as they work on projects for clients, the technical professionals remain outward-looking and keenly receptive to the innovation needs of industrial companies.

The second activity is the *spinning out* process, since Generics also acts to nurture and incubate new ventures. The contract R&D activity described above acts as a source for innovations. Occasionally, these innovations and their corresponding intellectual property rights may become the basis of a business plan championed by a small group of employees. Following a dialogue with management, their venture is refined, and further work supported by Generics is carried out in order to provide additional information. The Innovation Exploitation Board reviews the project and helps the team to build their business case. This board is a small group of about five to seven individuals representing a spectrum of the relevant business and technical areas. It meets weekly to review the candidates for spinning out and the investment opportunities.

When a business case seems reasonably promising, the innovation board gives the green light for a small group of professionals (two or three of them) to champion the venture to leave the laboratories and create a start-up company based on the venture. This is called a 'spin out'. Each year, several start-up companies are incorporated this way. A list of recent spin out companies is shown in Table 4.1. Specific examples are discussed below.

Charged with the task of growing their business, the team moves to another building, but remains on Generics' premises. The team thus benefits from access to, and considerable informal interaction with,

Table 4.1 Examples of start-up companies spun out by Generics

Year	Name	Description
1991	Sensopad Technologies Plc	Supplies components and technical expertise under patent licensing arrangements for innovative, low-cost, non-contact sensors and controllers.
1996	Flying Null Plc	Magnetic tagging technology permitting a gap between tag and reader head. Used to track goods and art.
1997	Imerge Plc	Hard-disk based audio and video software and devices for consumer, retail and corporate sectors.
1998	Synaptics Inc. (formerly Absolute Sensors Plc)	Non-contact positioning technology. Applications in automotive, game and pen input markets.
2000	quantumBEAM Plc	Free-space optical communication technology, transmitting bit rates up to 5 Gb/s. Fixed wireless technology.
2002	SPHERE Medical Plc	Set up to exploit a commercially Siemens developed silicon sensing technology for blood gas analysis.

Source: Company information and www.genericsgroup.com.

their former colleagues, whom they know well since they have worked closely with them over the previous years. This is extremely important, as the entrepreneurs usually are very lonely in their efforts. In the present case, they have access to a pool of expertise from former colleagues to help them progress their business. They benefit from informal contacts, suggestions and guidance on a range of topics.

Generics is the main shareholder of the newly formed company; the remaining (minority) shares are owned by the founding members. All the intellectual property rights are transferred to the start-up. The latter focuses its energies on growing the new business and on financing this growth. The business development is continually helped by advice and suggestions for contacts provided by the 'parent company' Generics. This phase of growth takes two to four years, but in recent years the duration has become shorter. When the business is at last well established, the priority becomes to find a buyer. Here again, Generics' extensive network of technology companies and investors is very useful. Eventually, the spin out company is sold in a *trade sale*, at a price representing a multiple of the amount invested in it. The proceeds of the sale go to the shareholders, primarily Generics, and they help finance the third activity, described below.

Some specific examples illustrating the spin out process are described below. Spin out companies that have emerged from the 'start-up factory' of Generics have become part of the company's extended family. Close contacts are maintained with the 'parent' company at the personal and business levels, especially since they fall in to the same general sphere of activity. These contacts also serve as antennae providing business intelligence on changes in the marketplace.

In addition to creating ventures from its own ideas, Generics occasionally partners other companies in order to help commercialize R&D projects carried out in their R&D units. One recent example is the Siemens Corporate Technology group in Munich. In this case, a project was identified by Generics and brought into the start-up company Sphere Medical Ltd., incorporated in September 2002. This company, now housed in Generics' incubator in Cambridge, aims to exploit Siemens' silicon sensing technology, which, among other applications, will be used for blood analysis. Another example of partnering was the 'Brightstar' incubator, described in the following chapter as an example of how British Telecom created value through spinning out start-ups from one of its corporate laboratories.

The third activity of Generics concerns *seed capital*. Generics manages a number of funds for investing in its own intellectual property and in spin out companies as well as in external start-ups in technologies and business areas familiar to Generics. Most of these are based on 'robust technology', according to 'Cambridgespeak', in the areas of strengths of the laboratory. In this way, the investor brings 'smart money', that is not only funds but also knowledge of the technology and of the business, which represent a plus for the start-up.

One such fund is the InterTechnologyFund (ITF). Formed in 1996 to date this fund has close to 30 investments in technology-based ventures. Another fund was started in Switzerland in 2000, ETeCH, aimed at helping commercialize innovations generated by research laboratories of universities. Generics also invests its own capital, managed by Generics Asset Management Plc., an internal company regulated by the UK financial services authority.

Each of these funds has its own management team searching for new investment opportunities and managing the investment portfolio, as well as restructuring deals when needed. They may also invest in building business cases for ventures to be spun out as start-up companies. In all, the valuation of the portfolio was close to £35 million in early 2003, in accordance with the rules of the British Venture Capital Association. Generics strongly believes in connecting directly, on a peer basis with scientists in university laboratories around the world. In this way, opportunities are frequently identified for development which had not occurred to the researchers concerned. All scientists and engineers have experienced the value of exploratory give and take with colleagues, or clients, which often results in a spark of invention. This kind of discovery through interaction is one form of *innovation mining*, and will be discussed in the following chapter.

The following examples illustrate some of the issues encountered along the path for innovation projects to emerge as a spin out companies. They concern the business of sensors and free-space optical telecommunications. Both involve strong science-based patent positions.

The Spin Out Company Absolute Sensors:
Leveraging Technology via Different Channels

In 1994, an elevator company approached Generics to solve the problem of precise positioning of an elevator. To ensure that it stopped exactly level with each floor, computer control systems were required in which precision sensing should be used. The challenge arose because of the dusty operating environment, where precise alignments and positioning would be difficult to achieve. In addition, the sensor would be subject to changing temperatures and, to give it a longer life span, must not have any contact with the moving elevator.

Generics developed a technology, which became known as Spiral. This new sensor functioned with magnetic induction, via an electronic board that generated an alternating current. The current produced a magnetic field that could be picked up by the sensor on the other side of the gap, causing it to resonate. The signal was then sent to the processing electronics and converted to a precise location and angle measurement. The Spiral technology was very attractive; not only was it non-contact and able to function in a dusty environment, but it could also be produced cheaply, was highly precise and could be used for measurements in two or three dimensions.

This sound, scientifically based invention was protected by a patent. After identifying the range of potential applications of this sensing technology and successfully obtaining two licence deals for it, Generics decided to form Absolute Sensors Limited on 1 July 1998. Ian Collins, whose group had developed the technology, David Ely, an engineer specialist in resonance systems, and Malcolm Burwell, an experienced technology professional, left Generics to become Absolute Sensors' first employees. With the equity initially held by Generics, the company eventually obtained £487,000 in seed funding from the Group and opened its capital to employees, with a portion of the shares for the three founding partners. The team moved to Generics' incubator, a renovated eighteenth-century mill, 50 metres away from the laboratory where they had worked for many years as Generics' employees.

By putting this innovation to work in different applications in sectors such as automotive, game and pen input devices – used, for example, in palm assistants – Absolute Sensors was soon able to raise additional capital (£460,000) from one of Generics' funds as well as from other investors. Finally, Generics sold its 80.3 per cent interest in the company to Synaptics Incorporated for cash and equity representing an aggregate value of $3.3 million in 1999. By the end of that year, Generics obtained a 2.7 per cent interest in the Californian firm Synaptics Incorporated, which became a public company in January 2002.

The Spin Out Company RETRO: Pitfalls in Collaborating with another Venture to Develop a Free-space Optical System

In early January 1998, Alan Green, Technical Leader of Generics' internally funded Intellectual Property (IP) development and exploitation team, Retro, had a conundrum on his hands. The technology at the core of their new business opportunity depended on the development of a semiconductor optical modulator, for which the team did not have the required expertise or infrastructure. The choice was between developing this capability internally, or finding and partnering with a non-competitive organization which had this technological competence.

Two of Generics' senior managers, key sponsors of the Retro technology, thought that the development should be done in-house, thereby providing full ownership and control of all the IP already existing and to be generated in the course of the future work. The 'in-house development' option was not, however, as simple as it seemed at first sight. Would the team be able to develop this new technology? How long would it take? Would the new technology be better than that available elsewhere?

Aware of these risks, Alan was inclined to pursue the second option. He was hoping to cut a technology development deal with Kelvin Nanotechnology (KNT). This research company, operated by the Optoelectronics Research Group at the University of Glasgow, had been at the forefront of nanoelectronic, optoelectronic and bio-electronic research for several years. But doing business with a university-owned company proved to be more complicated than he had anticipated.

Providers of local telecommunications systems are usually limited by the fact that the 'last mile' (the connection between the end customer and the exchange) has low bandwidth and is implemented by a fixed cable installation. Upgrading the 'last mile' is costly and time-consuming. Similarly, radio frequency solutions are less attractive by their limitations on bandwidth, diverging standards and a crowded frequency spectrum that requires costly operation licences.

Existing conventional free-space optical systems are point-to-point links, where the transmitting laser and optical receiver have to be carefully aligned. The need for careful alignment on initial installation and subsequent maintenance, coupled with high cost, has until now prevented this technology from becoming a viable alternative to existing communication technologies.

The new technology platform is a type of free-space optical communication system, referred to as retro-comms. The key benefits of retro-comms include simple installation, eye-safe system and operation in an unregulated spectrum. The key component is a modulated retro-reflector. Each communication channel has a laser that is pointed at the retro-reflector. The beam is reflected straight back to the laser, but with modulation imposed on it. Optics in the receiver split off the reflected beam to extract the data.

The key IP is embodied in the retro-reflective modulator, which acts like an electrically controlled mirror. The mirror either reflects or it does not, in accordance with the voltage applied. The target performance of the modulator is to switch the mirror at 100 Mbit/s. The retro-reflectivity depends upon another important IP in the accompanying optics in front of the modulator, called a telecentric lens. The modulator, made of aluminium–gallium–arsenide, consists of a large number of mirrors, which can be switched independently, so that the device can support a multitude of communications channels simultaneously.

The decision to partner or develop internally
The idea for development of a retro-reflective modulator had been around for some time. Organizations like Nortel, Lucent and GEC had done some work in this area. The stumbling blocks were the difficulties in obtaining and maintaining accurate alignment, and in achieving point-to-point connectivity at a reasonable cost.

The IP that enabled the Retro technology to overcome these problems resided in the optical system and the modulator physics. Generics already possessed the relevant optics design capability from previous client projects for development of range finders and similar products. The development of a modulator, however, required levels of knowledge and specialized equipment that only a leading-edge research organization could provide. When Alan Green completed a search to identify candidate partner organizations, he found that there were not many options.

The most suitable research partner he could find in the UK was Kelvin Nanotechnology (KNT). This company was the only 'one-stop shop' in the UK in this area. It had been created to facilitate the commercialization and exploitation of the world-class technology and expertise available. This group had been at the forefront of nanoelectronic, optoelectronic and bioelectronic research for a number of years, with research groups in nanoelectronics, millimetre-wave integrated circuits, optoelectronics, molecular beam epitaxial (MBE) growth, bioelectronics, silicon sensors, dry etching and plasma processing, device modelling and simulation.

Issues in the collaboration
Two issues which stood in the way of setting up this partnership became apparent during initial discussions in June 1997. The first, ownership of IP, was not unexpected, and Generics had much experience of similar negotiations with clients. The second, and perhaps more surprising, issue was found in the university terms of business.

The IP negotiations began in July 1997 with discussions about a confidentiality agreement. KNT rejected this confidentiality agreement, and discussions began as to how the partitioning of IP, necessary to resolve issues of protection and ownership, could be embodied in any agreement. The parties to the agreement would have to include the University of Glasgow as well as KNT and Generics.

More incredible was a clause in the Glasgow University terms of business, which would hold Retro (and therefore Generics) liable for any consequential loss to the university facilities occurring during the development of a prototype. It took several months to persuade Glasgow University to waive the consequential loss clause in their terms of business. Thereafter, and although it had taken from October 1997 to get to this point, in June 1999, the team produced their proof-of-principle modulator and it worked very well.

Alan would definitely have made the same decision again. The alternative under consideration was to employ a semiconductor-chip designer and use a commercial manufacturer to develop the technology, which would have been more expensive than the chosen route by an estimated factor of ten. And this would still not have guaranteed a speedier delivery as the designer would have been working in some isolation compared to the environment in KNT; the learning curve issues would also have had a big impact.

Developments move forward

The successful technology prototype gave the Retro team the opportunity to explore further applications for the core IP, and this resulted in nine other patent applications.

Once the first modulator had been built into a successful technology prototype, KNT was engaged to design and make another series of modulators. A semiconductor-chip designer and wafer-design software have since been added to the Retro team's capabilities, to continue development of the original designs produced through collaboration with KNT.

Andrew Parkes joined in June 2000 as CEO of a spin out company to lead the commercialization of the Retro technology. His first challenge was to identify appropriate partners to fund the spin out. He strongly felt that the partners chosen would have to add value in a way that an ordinary bank or venture capitalist would not be able to. Negotiations were started with Sandler Capital, who later brought in Intel Capital. Both these fundholders were keen to add optical communications ventures to their portfolios.

Intel Capital and Sandler Capital jointly took a $10 million position in the spin out. Generics retained a 59 per cent interest, and 20 per cent as reserved for the employees of Generics (as all staff are invited to participate in the exploitation of IP). The spin out now has a capital valuation of $45 million, and the quality of the new investors has led to substantial interest in further funding provision from a number of leading venture capitalists and fundholding institutions.

The Retro team has been growing rapidly, with new technologists and commercial managers joining, so that the spin out company has had to move out from the Generics 'incubator'. The market focus is still on the last mile, although some other interesting 'new technology, new market' opportunities have arisen. Test trials are in progress in Cambridge and the objective is to qualify the engineering status of the hardware and proceed with customer trials in mid-2003. In parallel, technical developments are taking place to build the Retro platform as a means of addressing additional markets.

These two examples of spin out companies are both high value-creating ventures. They highlight some of the issues raised when innovation projects are incubated into new businesses.

The first example underscores how in the course of the development of a venture income is generated by successively using different modes of commercializing technology: contract research, licensing out income and trade sale.

The second example highlights the perennial issue of the control of patent rights by a company when collaborating with another organization. It also cautions against unexpected clauses in policy rules of the other organization. This surprise factor caused a substantial and very detrimental delay in the progress of Retro, in an extremely fast-paced business area.

Generics' Business System

Generics can thus be described as an innovator/incubator/investor hybrid. This unique way of grouping different interdependent activities under the same roof is illustrated in Figure 4.1. The added value produced by this model results from the three complementary activities, which enable frequent interactions between them. The technical services/engineering laboratory is a window on the marketplace, with an extensive network of technology companies worldwide. This component also acts as a source of innovations, which will become the substance of the 'spin out' companies.

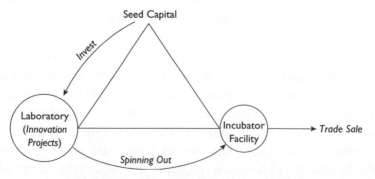

Figure 4.1 Generics' business system.

Finally, the returns on the investments generated by the trade sales provide resources for investing, on the one hand, in project teams developing their business plans prior to spinning out their ventures. On the other hand, Generics put seed capital (a couple of hundred thousand pounds per deal) in external ventures in order to build a portfolio of investments. Acting as venture capitalists is possible because the composite of Generics' activities supplies the competencies in technology and business necessary rapidly to carry out due diligence on the ventures. This integrated approach makes it possible efficiently to practise seed capital. The yearly turnover of Generics is as follows (in £ million):

- Sales of technical/advisory services: 17
- Licensing income: 1
- Investments: 8 to 12, depending on the year.

In brief, Generics' innovation model is based on the following characteristics:

- interdisciplinary work: once a problem is defined together with a customer, the approach to solving it involves a robust dialogue across technical disciplines, such as computer science, physics, chemistry, biology.

- free flow of communication: the layout of the office space emphasizes the importance of communication between employees: there are no private offices at Generics. Every staff member, including the CEO, works in an open-plan environment.

- risk-taking is central to the company's entrepreneurial culture. As Gordon Edge says, 'it is vital that the CEO of a technology company is seen to take risks, by being directly involved in an innovation project'.

This unusual set of characteristics demands unusual employees: Generics' staff comprises excellent scientists with high entrepreneurial energy and a keen business sense. Professionals who want to join Generics are very attracted by the idea that in joining they may well one day have the opportunity of launching their own spin out company in an environment that is fully supportive – because that is its *raison d'être*.

This exciting environment, however, has a price: each year a number of talented individuals leave Generics, as they form the spin out companies and grow their own business. The resulting loss of staff must be compensated by aggressive recruiting, especially in a place like Cambridge, where competition for talent is fierce. In Chapter 5, we will come back to Cambridge as 'Europe's Silicon Valley', a dynamic region for enterpreneurship and technology firms.[2]

Conclusion

In brief, the Generics model of having the unique combination of these three types of activities under the same roof, prompts two comments. First, the firm is in a position of activating the channel best suited to leverage most effectively technical innovation and intellectual property. These different channels are: technical services and contract research, licensing, partnership and co-development, incubating start up companies and investing.

Second, Generics' unusual profile attracts a rare breed of *researchers–entrepreneurs*. Indeed, this is a privileged place to work for scientifically trained individuals with a good business sense. Such candidates are seduced by the idea of having the possibility of starting their own company one day, while benefiting from a very supportive environment.

For these two reasons, it is argued that technology companies should look at Generics as a model to emulate, in order to generate revenues from technology and intellectual property through a variety of channels. This is advocated in the following chapter on the novel approach of distributed innovation.

Notes

1. Georges Haour et al., *Generics Case Study* IMD-3–1101 (June 2002). The Retro story is derived from a Generics document, written by Julian Fox.
2. Segal Quince Wicksteed Ltd, *The Cambridge Phenomenon: The Growth of High Technology Industry in a University Town* (Cambridge, 1998)

CHAPTER 5

Redefining Innovation Management: the Distributed Innovation System

During the last fifteen years, companies have extensively restructured their operations. They have bought and sold business segments to rearrange their portfolio of activities, the rationale being to concentrate resources on those activities in which they considered they had the best chances of winning in the competitive race. Firms have therefore drastically redefined their *business perimeter*. In some cases, this redefinition has been so radical that it has meant a *metamorphosis* of the business activities of the firm. In these relatively rare cases, the firm has voluntarily engaged in a relentless transformation of the nature of its business. Examples such as Danone, Nokia and Samsung will be described.

As part of this business restructuring, manufacturing companies have redefined their *production perimeter*. In an effort to decrease costs and increase flexibility, they have outsourced a growing part of their production. To a great extent, however, the development of innovations has remained within the firm. In order to turbo-charge the effectiveness of the innovation process, technology firms must now break away from internal innovation development and redefine their *innovation perimeter*, as they did for manufacturing. This new approach means opening up the innovation process considerably in order to cooperate with different external actors. In Chapter 4, the unique business model of the company Generics was discussed as an example pointing the way forward in this area. It is now argued that, following this model, technology companies should draw on a multiplicity of ways to generate revenues from their technical resources. This must be done by considerably stepping up interaction between the firm and external contributors. The firm and its external partners constitute the *distributed innovation system.*

The objective of this new approach to innovation is twofold. First, the aim is to generate value by effectively commercializing the company's technology by acting through different channels; this will be discussed in this chapter. The other objective is proactively to tap external technical resources in order to integrate them into a wider innovation system, choreographed by the firm. The firm must reach for those technical resources which are most appropriate for the development of new products and services considered to be the key for its future growth; this will be discussed in the following chapter.

Redrawing the Company Perimeter: Danone, Nokia and Samsung

In order to adapt to a fast-changing competitive environment, companies are continuously restructuring their business activities. In a small number of instances they go as far as radically transforming their businesses. Redefining their activities also implies considerable changes in the manufacturing operations. A similar change must take place in the way the company's technical expertise is converted into products and revenues in the marketplace.

The Danone Case

One example of a complete business transformation is the French company Danone, which, under the leadership of Antoine Riboud, drove the company BSN, a glass-making company founded in the eighteenth century, to go from manufacturing glass containers to becoming a main player in the food business – yogurt, mineral water, biscuits. This dramatic switch of business *from container to content* was achieved over a period of three years in the early 1970s through a series of acquisitions and divestments. Leadership and tenacity were certainly key elements for this success. A very supportive bank was another important asset in carrying out this considerable reshuffling of business activities. The sales volume of Danone was 14.3 billion € in 2002.

The Nokia Metamorphosis

Another example of business metamorphosis is the company Nokia in Finland. This company, founded in 1865, was originally in the forest industry. Later it became active in the production of telephone and power cables, as well as rubber boots, which were mostly exported to the USSR. The recession following the 1973 oil crisis convinced Kari Karaimo, who became managing director in 1977, that Nokia's business had to change radically. Through tireless efforts, he imposed his vision of turning Nokia into an electronics giant. The extensive selling-off of commodity activities and the buying of equity in telecommunications companies such as Mobira and Salora were continued by Karaimo's successor, after his death in 1988.

This transformation took place within the context of a country undergoing drastic change and opening up to the world. For this reason, Finland has been called 'Nordic Japan'. The present CEO Jorma Ollila was appointed in 1992. Under his tenure, major investments have been made to build the Nokia brand, and the importance of design has been recognized. The remarkable journey of this titanic restructuring is now serving as a background leading to preparations for future business changes at Nokia Venture, to be discussed in Chapter 6, as well as to a disciplined management of innovation and of the supply chain. In 2002, Nokia's business volume was 30 billion €. More than 75 per cent of this was from the mobile phone division; the balance was essentially the networks division, with Nokia Venture representing less than half a billion €. The overall turnover of Nokia alone represents close to 6 per cent of Finland's GDP.

The Saga of Samsung Electronics

Now a recognized leader in the electronics industry, Korea's Samsung started in the business of trading food. In 1936 the Chairman Byung-Chull Lee opened a rice mill in Masan, Korea. The business was successful and he was able to buy first a transportation company, and then a real estate business. In 1938, he created Samsung, expanding its food business by exporting fresh produce to Northern China and Beijing. Relocated to Seoul, Samsung became a large trading company and after the end of the Korean war diversified its activities into manufacturing, founding Samsung Electronics in 1969 to produce black-and-white television sets.

In order to acquire the knowledge necessary to build a strong electronics business, Samsung established a joint venture with Japan's Sanyo. This consisted of the assembly in Korea of electronic components made in Japan, a country which was very much a role model for Korea at the time. The next milestone decision was when, in 1977, Samsung bought Korean Semiconductor from its founder, a company that produced transistors and integrated circuits for watches and home appliances.

Building on President Park's policy of developing key export industries and, convinced by the economic problems triggered by the oil crisis of the 1970s, Chairman Lee decided that Samsung had to move up the value chain and enter a particular segment of the semiconductor industry, the DRAM – dynamic random access memory – chips. This decision was announced in 1983. Semiconductors became a key activity of the Samsung Group, by then one of Korea's largest conglomerates, or 'chaebols'. This move opened a new phase of acquiring and learning technology. It took ten years for Samsung to become the largest producer of memory chips in the world. It is still number one today.

A first step was to again follow Japan's example. The 1980s were buoyant times for the Japanese economy, and its electronics business in particular. Sanyo was a privileged partner in this informal 'mentoring'. In addition, steps taken by Samsung included:

  hiring US-educated engineers.

  establishing task forces to elaborate the blueprints for the development of the DRAM business. These teams were often duplicated, one in Kiheung, one in Silicon Valley, California, in an attempt to enhance the quality and the quantity of input, as well as to provide a diversity of perspectives on development. This was particularly the case for the more advanced generations of chips. On several occasions, the Kiheung team came up with the winning solution selected by Samsung.

  securing an agreement with Micron, Idaho, to obtain the 64K DRAM technology.

The agreement with Micron, however, did not cover critical details of the technology. This was in conformity with the 'export control' rules put in place at the time by the USA to protect their technology. Samsung then realized that it had to develop its own design and manufacturing technology. It met this challenge brilliantly; a prototype of the new chip was developed in a record six months. Dr Sang-Joon Lee, leader of the development team at the time, recalls: 'I was so immersed in the work, I stopped smoking and drinking; I hardly slept more than four hours a day for six months.'[1]

Using lithography machinery from Japan, the production plant was built and commissioned, also in record time, mobilizing an astounding level of commitment from Samsung employees as well as from suppliers. As Lee recollects: 'We started building the chip manufacturing plant in September 1983. We completed it in March 1984. I heard that it took 18 months in foreign companies. Experts from Intel, IBM and Japan were very surprised at our fast construction and successful test running of the plant.'[2]

Maintaining the momentum of total commitment, Kun-Hee Lee, Chairman of Samsung Electronics, succeeded his father in 1987. He said: 'Samsung is many things to many people: Innovator, Economic power, Partner, Survivor, Employer, Helper, Leader. But more important than what we are to people, is what we do for them.'[3] He further strengthened Samsung as a technology leader, acquiring firms that brought relevant technologies, such as Harris Microwave Semiconductors. He also divested ten of the group's subsidiaries to sharpen its focus on electronics and engineering. Samsung is now one of the top ten recipients of patents in the USA. The current CEO, appointed in 1996, is Yun Jong Yong, who was heading the company's operations in Japan. He is an engineer by background, and maintains that some degree of chaos is necessary to keep Samsung agile. He had to cut costs drastically during the 1997–8 Asia crisis. An article in *Fortune*, dated 24 January 2000, said of him: 'This creative, gutsy man swiftly introduced a stream of innovative high-tech products.'

In 2002, Samsung had a $25 billion sales volume. Its profitability of 17 per cent of sales was much higher than that of its larger competitor Sony, with only 1.5 per cent profit of sales. Today, the 'boom and bust' chip business accounts for only 23 per cent of Samsung's total sales. The company remains a pacesetter in the fields of memory chips and thin-film displays, as well as a leader in many consumer electronics products. It includes mobile phones, for which the company came from nowhere to its current number four position worldwide; it is one of the most widely recognized international brands. It produces the winning ultra-thin SENSQ 760 laptop for Dell, among others. Quite a road has been traveled since the rice mill of Masan.

The examples of Danone, Nokia and Samsung represent extreme cases of the considerable restructuring that is continually taking place in business. They also highlight how important technology acquisitions have been in their metamorphosis process, particularly in the case of Samsung. The examples will be considered from this perspective later.

The restructuring of the companies' business involves considerable buying and selling segments of activity. In the recent past, General Electric (GE) was a classic example of such reshuffling; its objective was to retain within the conglomerate only the top performing activities, that is, those ranking number one, two or three in terms of market share in their categories. Interestingly, there are occasional exceptions to this rule: GE has kept control of the television network NBC, although this non-technical activity does not satisfy the requirement of having a leading position in its market. Management is not an exact science.

One aspect of this restructuring involves a trend towards outsourcing manufacturing activities. This externalization reduces the inventory and the amount of capital immobilized, and also provides greater flexibility. Following the example of Japanese companies, the car industry first adopted this approach, essentially retaining the assembly, distribution and sales, as well as the brand building. Suppliers not only deliver on a *just in time* basis, but are also invited to assemble components themselves, for example car seats, on the automotive assembly line. In another recent example, Ericsson in 2000 decided to farm out the manufacturing of mobile phones to Flextronics. Currently, this US/Singaporean company assembles mobiles for three of the main competitors in the industry: Nokia, SonyEricsson and Motorola.

Outsourcing all, or portions of, the production activities raises specific management challenges in order to smoothly integrate the input along the supply chain. Fully outsourced production has been the business model of computer maker Dell, now the largest producer of personal computers, before HP-Compaq. This has led to the somewhat bizarre expression of the *virtual company*: the fact that transactions between firms are facilitated by ICT – information and communication technologies – does not mean that the firms are any less real than when the telephone is used for business dealings.

If technology companies have reorganized their manufacturing activities to redefine their production perimeter, is it not time to consider a similar overhaul of their *innovation perimeter*? In rethinking the way they envisage innovation, technology firms will associate a number of external contributors with their innovation process. The business model of Generics provides an inspiration for such firms to proceed along this path. They must increasingly choreograph a diverse array of channels for commercializing technology. These channels are illustrated in Figure 5.1, and described below.

In the distributed innovation system, firm *A* is situated at the hub of a network comprising company, government or university laboratories, contract research and advisory service organizations. The all-purpose word 'network' does not convey the notion that *proactively*, continually, and in a coordinated way, the hub-company A looks for ways to maximize revenues from its pool of technical expertise. The new concept of distributed innovation system has been coined in order to convey the notion of activating contributors in this network, with a view to identifying, preparing and concluding transactions in a coordinated way. Just like Generics, the company identifies

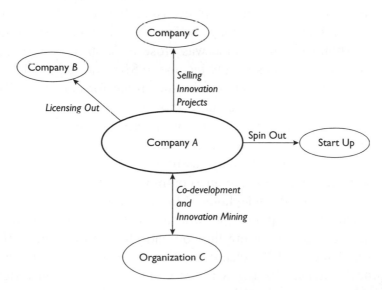

Figure 5.1 Proactive leveraging of technology in the distributed innovation system.

and actively looks at each item of technology as well as the most appropriate route to commercialize each element of its technology, while balancing and maximizing the value-creation and taking into account the strategic interest of the firm. These different routes are described below.

Licensing is a traditional mode of creating revenues from technical innovations. A *licensor* company grants another, the *licensee*, the right to use an innovation, based on specific intellectual property and know-how. Patents represent the leading basis for licence agreements. The licensing trade represented a volume of $142 billion in 2000 worldwide; of this, licences on pharmaceutical products constituted the lion's share of $10.5 billion. This represents a considerable increase from 1990 when the licensing volume was only $10 billion. These numbers underscore the very substantial increase in technology flows worldwide.

At the company level, securing revenues from licensing requires a full commitment and willingness on the part of top management actively to exploit this source. Its implementation demands that the company organize an effective team charged with maintaining their network up to date and activating it to identify prospects. Skills for negotiating and settling contracts are then required to conclude licensing deals based on the firm's portfolio of intellectual property. For example, IBM and Thomson have been very active increasing their revenues with royalties from licensing. Between 1990 and 2001, IBM increased its licensing income from $30 million to $1.9 billion. As for Thomson, this income was close to $500 million in 2002. Dupont also boosted its royalty income to $100 million in 2000.

Co-development constitutes another route for commercializing technology. It associates the hub-company *A* with a partner, in order to carry out a collaborative innovation project. The partners pool their resources: technical competencies, market intelligence and capital. Co-development may involve all companies active in an industry, for example to develop a common standard such as the GSM standard in mobile telephony. Manufacturers of handsets and equipment, their suppliers, as well as operators, are stakeholders working together to ensure the compatibility and openness of the various elements of the chain; an example of this is the *Bluetooth* approach for establishing wireless connections between different equipment such as personal computers and cellular phones.

This allows the sharing of development costs and risks. More importantly, it is likely to enhance the effectiveness, that is the market relevance, of the innovation, by involving the customers. It may shorten the 'time to market' as well. For these reasons, oil companies have for a long time formed consortia to do the prospecting and exploiting of oil and gas fields.

Examples of co-developments are numerous. Japan's approach to consortia involves a large number of companies and institutions, under the leadership of one company. These consortia are facilitated by the METI – Ministry for Industry (formerly MITI) – with various degrees of success. A number of such consortia are currently under way on themes as diverse as: supraconducting ceramics or bio computers. For each consortium, each company seconds staff to the project in which it participates. Participating companies share the intellectual property developed in the course of the project. When the collaboration ends, the partners are free to leverage whatever the project has generated as best they can to achieve commercial success. This cooperation among competitors included a dimension which caught the attention of Western observers in the late 1970s. Since then, similar initiatives have been launched in the USA.

Responding to the challenge from Japan, the USA shifted its policy considerably in the 1980s, in order to remove the obstacle represented by Federal antitrust laws to collaborative projects. The resulting 1984 National Cooperative Research Act (NCRA) paved the way for the creation of the Sematech consortium, established in 1987 to group 14 US companies from the semiconductor industry, including HP-Compaq, IBM, Intel, Lucent, Motorola. The objective was to develop processing and materials improvements for the manufacturing of advanced semiconductor products, with a mid- to short-term horizon of three to five years. Federal Defence funding represented the relatively modest sum of $80 million per year during the 1987–96 period; since 1997, the funding has been entirely private. Participating companies commit themselves not to engage in any agreement that would restrict output or capacity of technical innovation. At the same time, the Advanced Technology Programmes (ATP) began to provide Federal seed funding (roughly $130 million in 2002) to consortia comprising companies and universities or government laboratories. Public funding was matched by private sector support. To date, more than 500 projects have been supported, representing well over $3 billion of cumulated effort.

Precursors of this kind of initiative in the USA include the National Science Foundation's industry–university cooperative research centres programme, as well as the Palo Alto-based research organization EPRI – Electrical Power Research Institute, supported by electricity utility companies.

Equally eager to capture the 'spill over' benefits of co-developments (that is, a participating firm uses the know-how generated by another, which would otherwise be lost), the European Union (EU) has carried out collaborative development programmes for 20 years. These so-called Framework Programmes did not run into the obstacle of anti-trust laws, since Europe does not have strong legislation in this area compared to the USA, and it also has a stronger institutional bias towards trying to reach consensus on technical standards, such as for high definition television, GSM and, currently, the zigbee standard for wireless, involving some thirty partners.

The main objective of the Framework Programmes is to catalyze collaboration across Europe. The projects involved are very diverse. They may concern materials for the automotive industry, the development of new technologies in the telecommunications or fine chemicals industries. In this type of joint project, a group of very diverse partners is established to carry out a specific development. Participants in one project may include small and large companies, public laboratories or universities from many different countries. It is a multinational, multicultural effort involving partners of different sizes, both public and private.

Such consortia projects are thus characterized by a maximum degree of complexity. The challenge of managing a group of, say, fifteen diverse partners, who may be located from Greece to Finland, is unique in the world. Too often partners rush to cluster together in order to profit from a funding opportunity. They do not spend enough preparation time in dialogue and defining how best to leverage their complementary capabilities to achieve the project objective. Not surprisingly, this insufficient alignment of the various participants results in problems later.

The Framework Programmes of the European Union represent a small percentage (around 4 per cent) of the overall R&D investments, public and private, made by the fifteen countries of the European Union. In spite of the relatively small financial impact, it provides a powerful incentive for the various actors to collaborate across the borders. The commitment demonstrated by the non-EU country of Switzerland to participate in these Framework Programmes may be taken as an indication of the worth of their contribution.

The sequencing of the human genome is another type of consortium. It was completed in the spring of 2003, at the time of the fiftieth anniversary of the description of the structure of the DNA, in Cambridge. This consortium involves teams from the USA, Great Britain, France, Germany, Japan and China. The objective was to map the gene sequences with the use of public money, in order to provide a basis of information that would be generally available to all for medical and pharmaceutical developments.

Selling innovation projects is much less frequently practised. Companies frequently buy and sell portions of their business activities in order to readjust their activities, but they rarely do this when it comes to buying or selling innovation projects.

Firms frequently discontinue development projects. This may be because the anticipated economic potential turns out to be insufficiently attractive or because the firm has changed its business strategy. Support for a project may drop when the manager championing it leaves the company. Whatever the reason, it is a painful step: human organizations do not like to abort a venture. As a result, they tend to procrastinate and not manage the interruption well.

This induces unnecessary confusion and demotivation of the staff. Management must make the effort of fully explaining to the team members why their project is being interrupted. This must be done in a trusting face-to-face dialogue with the team; the temptation to announce and explain such a decision by a memo or an e-mail should absolutely be resisted. This is definitely a case where managing by memo is not enough. By seizing the opportunity to engage in a dialogue with the project team, management will have the chance to maintain their motivation level, while explaining the sound business rationale for the decision. The point is to disconnect the project from the members of the team: the idea is be to 'kill' the project, not the team.

It is partly because companies feel so uneasy about discontinuing projects that they are inclined to turn the page and consider the investment in a project as a sunk cost. Instead, the firm should ask: if this project is not right for us, what can we do to create revenue from some, or all, of it? Who could be interested in buying it, to help us recoup our investments in this development? Companies usually do not even raise these questions. It is probable, though, that the firm knows which other companies might be interested buyers. These are likely to include firms that are their current customers, but the circle of potential buyers is wider than that and they could be found anywhere in the world.

Identifying potential buyers for a given project requires diligent homework, based on an excellent and current knowledge of the industry. It needs an outward-oriented mindset, particularly on the part of the R&D knowledge workers, who must routinely monitor intelligence regarding the industry in the course of their work. This imaginative quest aims at identifying the firm for which the particular unit of know-how embodied in the project will represent the highest value. Selling an innovation project represents a challenge of a very different nature than selling a product. That said, however, there are precautions to take; it might indeed be ill-advised to enter into negotiations with competitors if this were to result in leaking sensitive business and technical intelligence to them.

In any case, the decision to sell, and the search for a buyer, must be undertaken promptly, as the value of the technology will rapidly decrease. The 'shelf-life' of technical developments is very short, especially in fast-paced sectors, such as ICT – information and computer technologies. Furthermore, in the case of patents, the project value decreases as the diminishing remaining life of the patent lowers the attractiveness of a potential licensing agreement based on this patent. Examples of selling innovation projects in two different industries are given below.

An industrial gas supplier had worked on the development of an innovative, patented chemical process for a number of years. The cumulated investments in the development totalled 40 million €. For sound strategic reasons, the project was interrupted; the project team was disbanded and the pilot plant was left to rust in a hangar. At no point did the company ask whether the project could be sold as a complete package including know-how, intellectual property rights and prototype. Potential buyers were known to the firm, since they were chemical companies already buying industrial gases from it. To sell would not have meant leaking proprietary information to a competitor, yet no effort was made to explore the interest of possible buyers.

Another example concerns a laser technology under development at Philips' Natlab Corporate R&D establishment. At some point it became clear that the ultimate business would be too small and too far removed from the firm's key business areas. Philips' executive committee therefore decided to sell it. After some analyses and preparatory work, several buyers were identified. As it turned out, this specific bit of technology was found to add considerably to the buyer's technology base. As a result, Philips was able to obtain a selling price that was a multiple of what they had originally expected.

Companies currently buy and sell business segments. In some cases, the technical component of these segments is a particularly important reason for the transaction. The internet and telecommunications provider Cisco has built its business on the acquisition and very rapid integration of technology-intensive segments which enhance its product range. Alternatively, it may go into a temporary partnership with firms, following the same objective. This way of extensively tapping into external young technology-intensive firms is in contrast to the much more internal approach to developments adopted by Lucent Technologies. The main reason for this is that Lucent has within its own walls the biggest part of the science and technology powerhouse Bell Laboratories, inherited from the time of the break-up of AT&T. The newer company, Cisco, did not have that legacy and thus entered the field with a very different mode of operation, relying heavily on acquisition of companies, which allowed Cisco to quickly expand its offer.

The practice of buying and selling innovation projects is bound to become much more widespread, as it allows firms to mine a new source of revenues that would otherwise not be exploited. It requires the proper mindset of continuous looking outside for possible buyers of technology; intelligence on potential buyers must be constantly gathered and updated. Protection of business-sensitive information must be ensured. Teams must be well trained to negotiate the technical, financial and legal aspects of transactions. Trusting lines of communication must be maintained between the management and the team working on the innovation project. As such a transaction raises the issue of the fate of the project team, this issue should be resolved through a trusting dialogue with the team. If a number of project members do not want to change employer when the project is sold, a technology transfer arrangement must be found.

Spinning out ventures is another way to create value in the distributed innovation system; selected R&D projects are spun out into start-ups à la Generics. An example of transforming laboratory projects into companies is provided by British Telecom (BT), which in 2000 established the 'Brightstar' incubator on the premises of the company's 2500 R&D staff at Adastral Park, in Ipswich, Suffolk, UK. The rationale was to create value from the large patent portfolio that BT has. It was expected that along the way this would also boost BT's share price. This initiative was, in fact, the result of a partnership with Generics, discussed in Chapter 4. Each partner was represented by members on a council, which assessed and selected the projects and their teams from the R&D unit. The selection council team evaluated their potential, with somewhat the perspective of a venture capitalist reviewing the presentation of a business plan. The selection team then gave the green light to those projects teams, which could move into a dedicated 'incubator' building, located on the premises of the laboratory site. This process was similar to that followed by Generics' Innovation Board. Figure 5.2 illustrates the

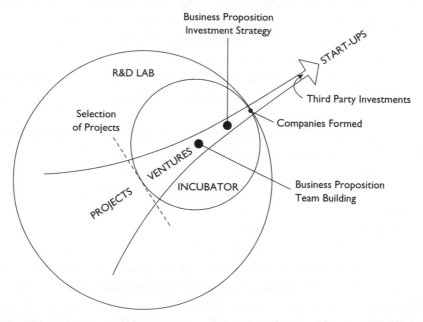

Figure 5.2 Schematic of the incubation process aimed at turning technical innovation projects into start-ups. A crucial element in that process is the venture coaching provided to the teams in the course of this transition.[4]

sequence of steps involved in this incubation process, from selecting the ventures, formulating the value proposition and business case, preparing a business plan, creating the company and looking for investors.

The process of turning an R&D project into an incorporated company took 18 months on average. During this period, each project team was coached by an experienced professional who accompanied the transition from project all the way to firm creation. This *venture coaching* activity is critical in helping turn engineers into entrepreneurs. One of the tasks of the coach is to ask the team difficult business questions in a friendly, but ruthless, manner, as a sanity check. He or she provides insights in assessing and sharpening the value proposition and the business model of the incubated company. The coach does *just in time* management development, which includes tutoring the team on specific business issues such as marketing strategy, intellectual property rights and the preparation of a business plan. The coach calls on external contributors when specialized knowledge is required, on legal issues for example. Coaching work also involves identifying gaps in the team's capabilities, recommending new hires, constantly sharpening the living document which is the business plan. It also involves making recommendations on possible investors, while at the same time helping the team to prepare a convincing presentation of their business case to the venture capitalists who are under constant pressure.

In the case of 'Brightstar', close to £15 million of non-BT financing was secured for the incorporated companies at the end of the 18 month period. BT retained substantial equity in these young companies. If among these ventures one ultimately develops into a success story with high market valuation on the stock exchange, this will be more than enough to bring handsome returns on the total investments made in the whole incubating process.

Spinning out start-ups from a parent company raises a number of issues. Among them are the following:

Intellectual property rights (IPR): parent companies tend to retain the IPR ownership, only granting a licence to the start-up. This is counterproductive. The IPR should be fully devolved to the start-up, so that no strings are attached at the time when external investments are sought. Investors want to participate in ventures in which there is a clear-cut situation avoiding any possible restrictions or conflicts regarding intellectual property in the firm's future activities. Imagine a situation in which a patent is filed, but depends on another a priori patent, for which the parent has retained ownership. The resulting contractual tangle may well cost the life of the young company.

Strategic risk: the parent company should indeed only consider spinning out activities that will not threaten their own most important business positions in any substantial way, when making a sensitive fraction of its assets accessible to competitors. This risk is often over-estimated, and rather than sitting on an underfunded development project, it is often much better to have a competitor provide the necessary investment for its development, even if both competitors will then benefit. In the case of BT, an example of this kind was a software which helped operators efficiently plan and develop green field wire or fibre networks.

Equity level: the parent company should not retain an excessively high participation in the start-up. External investors will be discouraged from investing in a venture if the reduced equity for sale restricts the scope for decision and control.

Management meddling: once the decision has been made which project to select for incubation, the parent company often succumbs to the temptation of intervening and interfering with its management excessively. Where needed, the role of the parent company is to provide positive support and guidance, not to micro-manage the project and then the start-up. It should provide a clear framework with sensible rules and then let go and allow the venture to find its own business trajectory.

Such corporate incubators generate value by creating equity in the start-ups for the parent company. They also have significant secondary effects. First, such an incubator, attached to an R&D unit, acts as role model infusing an entrepreneurial, business-oriented culture in that unit. In this respect, it should be mentioned that corporations often underestimate the readiness of their R&D staff to respond to opportunities for forming ventures. Soon after the creation of an incubator is announced, many more candidate projects for spinning out than anticipated are submitted. This was definitely the case for 'Brightstar', and it is also true in many other instances. This self-starting quality is extremely encouraging, as it mirrors the excitement of start-up companies. It also makes the management development work for the coach particularly rewarding. Second, the creation and operation of such incubators will boost the parent company's image as agile and exciting. This attractive image will in turn have the positive effect of attracting staff with high entrepreneurial spirit.

Spinning ventures out of universities: Instead of a corporate laboratory, the source of innovations may be a university research department. Technology firms may leverage parts of their own technology by fusing them with such start-ups, in what is called innovation mining, described below.

In the case of university ventures, the gap between the research project and the creation of a start-up is even greater, since the academic world is usually less concerned with markets and business issues. In this situation, it is even more critical to provide effective coaching to the project teams. Doctoral and post-doctoral students represent a population, which should be interested in the possibilities of creating a spin-out company, based on the work they are doing. This population is indeed the most likely to create innovations that might be candidates for conversion into business activities. A university entrepreneurship centre should thus focus on that population, maintaining a constant and informal contact with possible candidates–entrepreneurs. Such a centre should also organize a forum for discussions and workshops on issues such as intellectual property and patents, the process of creating companies and the preparation of business plans. This will expose researchers to business perspectives in a tailored manner and help them to see the commercial implications of their work. This will in turn alert them to not give away the results – and patent rights – of their research activities too lightly to any company. It is not unusual that university researchers are so happy to have projects financed by a company, pharmaceutical or otherwise, that they contractually abandon the patent rights to that company. In this way, the company acquires inexpensive innovations on which it can later build lucrative business segments.

Industrialized countries are putting universities under increasing pressure to perform research activities directly applicable to the private sector. Universities are thus strongly encouraged to act as a source of start-ups, following Stanford University's role model as a nursery for a number of young companies in Silicon Valley. Since the 1960s, the University of Cambridge has played a similar role in developing what is called the 'Cambridge Phenomenon', mentioned in

Chapter 4. This resulted in the creation of some 600 technology companies in 2003, employing 20 000 persons. Such companies are usually small. They rely on 'robust', patented technologies, such as biotechnology, medical devices, scientific instruments, sensors or new approaches serving the life-sciences sector, such as biochips and high-throughput screening. This close-knit array of diversified, agile, technology companies provides the Cambridge region with a resilience to economic downturns, apparently higher than the more 'boom and bust' Silicon Valley ecosystem.

The success of the Cambridge region in creating value is due to a number of factors. One is certainly the quality of talent and research carried out at the university. This provides a source of graduates and of specialized knowledge available to the start-ups. It also provides a 'brand' umbrella, which attracts young technology firms. These settle in the region, not because they have any prior special link with the university – alumnus or ongoing collaboration, for example – but to be in a position to benefit from the assets of the area – experts from the university, suppliers, subcontractors, specialized legal firms. In the case of Cambridge, there is another specific favourable factor: it is the *ethos* of trust particular to this university which diffuses to the start-ups in the region. This trust is manifested in the consulting policy of the university; there is no limit on the time which faculty members may want to devote to consulting. They are trusted to do their university job and not let their private professional activities interfere with that job.

Cambridge University is also generous with the patent rights: it yields the full ownership of rights, negotiated on a case by case basis, to the faculty or researcher concerned. This policy is now being reassessed, as the university wishes to generate additional revenues by controlling intellectual property rights, as most universities do. It is clear that these two characteristics have hugely favoured the transfer of know-how from the university to technology companies, as well as the remarkable development of the Cambridge Phenomenon.

The pressure of governments on universities to be more active in commercializing their technological innovations has meant that they have needed to build new bridges to commercial firms. This is done by seeking contract research funded by such firms. It is also done by encouraging informal contacts through geographical proximity; many science parks have been developed near university campuses in the

Netherlands, Sweden and the United Kingdom. It has also been done by changing elements in the legal environment; for example, the 'Law on Innovation', enacted in 1999, makes it now possible for public servants in France who are doing research in government laboratories, such as universities, CNRS or Inserm, to manage, advise or to be on the boards of technology start-ups. This has given strong impetus to a movement towards exploiting the large pool of excellent research carried out in these public institutions.

This should not be a threat to the 'curiosity-driven' research. As discussed in Chapter 2, this type of research must be continued. It is fully compatible, however, with more applied work going on alongside; there is plenty of scope for building many more bridges between public R&D and the private sector. To do this, public R&D institutions and their staff must be truly committed to commercializing their innovations. If this is the case, private firms will, no doubt, show more interest in the activities of public R&D than they currently do.

Another example of a law aimed to promote technology transfer from universities to the private sector is the Bay–Dole Act in the USA. This law was enacted in 1981. It was triggered by the fact that in 1980 only 5 per cent of the 28 000 patents generated at US universities, but held by the Federal Government, reached effective commercial applications. This very low rate of licences to industry resulted from the policy that the Federal Government retained the title – the ownership – of the patents and then granted non-exclusive licences to interested third parties. As indicated earlier, private investors are reluctant to participate in projects in which the intellectual property does not fully reside with the invested organization, the university in this case. Recognizing the problem, the Bay–Dole Act granted universities full ownership of the inventions made under federal funding.

Giving universities complete control over IPR triggered important changes. First, universities began filing many more patents: more than 2000 patent applications were filed in 1998, compared to fewer than 250 per year before the law was enacted. Among US universities, the University of California was the top earner of royalty income in 2000, with $261 million. These new revenues are invested in new research facilities, as well as in work towards filing patents and assisting candidates–entrepreneurs. For most universities, however, the added revenues do not cover the additional costs for patent filing and technology commercialization. Second, the law has resulted in a substantially increased flow of technology from university to industry. This has meant the creation of an estimated 260 0000 new jobs in the year 2000 alone.[5] The Federal Government should, however, closely monitor the patent policies of the universities, so that excessively broad patents or too exclusive licensing contracts do not restrict the social returns to society of publicly funded research.

In a similar way, Japan is currently making efforts to stimulate the currently low level of technology transfer from universities to the private sector. To this effect, a five-year government-funded programme was launched in 2002. This programme aims at creating technology transfer offices that will help convert universities' research into revenue-generating activities: contract research, but mainly licensing and creating spin out start-ups. A serious bottleneck in this area, however, is Japan's lack of experienced personnel able to carry out such activities, particularly selecting the projects and coaching the teams all the way to the creation of new companies. An additional handicap is Japan's underdeveloped sector investing in technology start-ups. It should be noted that a venture capital industry is emerging in Japan. There is also an apparent movement in which engineers are increasingly reluctant to join large companies and prefer to take the risk of creating their own firms instead. As in other first world countries, the collapse of the exuberant internet 'technology bubble' has put a temporary clamp on this trend, but a steady evolution is expected to continue in this area in Japan.

Innovation mining is yet another method to create value from technological innovation. In this approach, the firm seeks out a technology complementary to its own, so that the resulting ensemble presents a much higher value than the separate components.[6] This combined technology may be developed within the firm or in a joint venture. It may also be spun out as a separate project, using the venture trajectory described earlier in Figure 5.2. Examples include a prototype for a biochip, which was associated with a diagnostic tool. Another example is an optical device associated with a special chip for modulating light, in the Retro venture described in Chapter 4. Here again, the objective is to follow a path that will deliver the maximum overall value for the firm.

To make innovation mining happen, the owner of a unit of technology to be commercialized must have a clear vision of the business potential and an intimate knowledge of what may represent a value-adding complement. In this case the company's management displays a creative imagination, backed up by strong capabilities in scanning and gathering intelligence in the technological and commercial scenes. Alternatively, the firm may elect to call for specialized help by hiring an external service provider, which, acting as a *technology broker*, brings its experience and specific network in the industry. The services of technology brokers are compensated with a fixed fee, but also, and increasingly, with a success fee calculated as a percentage of the income generated for the client.

Conclusion

In brief, among the different channels of the distributed innovation system described above, technology companies make use of a small number of them – licensing and alliances. This is done in a piecemeal and *ad hoc* fashion. It is as if they had a piano, but only played a few keys. Realizing the value of technology must be done in a much more systematic way: the firm must play the full keyboard. Step up licensing activities, as IBM has done, co-develop, but also, sell certain innovation projects, spin out ventures and become involved in innovation mining.

For this, the CEO must consistently stimulate the technology-to-value process in the perspective of the distributed innovation system. A small group for commercializing technology should be formed to focus on this priority issue. The group should be chaired by the CEO and include senior technology, marketing, patent/legal and strategy executives. By choosing the most appropriate channel for each piece of technology, this group will generate handsome additional revenues, as well as more effectively leveraging the firm's technical development activities.

Companies must learn how to participate better and more consistently in the growing market for technology. In doing so, they will generate substantial, additional revenues. They will also continuously sharpen their intelligence on technical activities outside the firm, identifying opportunities and providing a way to calibrate their own activities as well. This practice will encourage their scientific and engineering professionals to be more outwardly oriented and business-minded. Such orientations will be increasingly crucial for future success.

Notes

1. G. Haour, *Samsung IMD 3-841 Case Studies* (2001).
2. *Ibid.*
3. *Ibid.*
4. G. Haour, *Incubating Technology Ventures: a shortcut to value creation?*, IMD Perspectives for Managers, no. 81(May 2001).
5. Technology Transfer. May 7 2003 report to the US Government Accounting Office. See www.congr.edu/files/publications_intellectual.
6. See, for example, www.idvector.com.

CHAPTER 6

Energizing the Distributed Innovation System with Entrepreneurship

In the previous chapter it was shown that technology firms must generate additional revenues from their technical pool of expertise. They need to convert their existing technological resources into income by 'exporting' some of them. To do this, they engage with participants of the distributed innovation system. But this is not all. These participants also constitute external sources of technology. Technology companies must be proactive in reaching out for these sources and 'import' some of their technical expertise in order to complement their internal capabilities. Channelling additional expertise in this way allows the company to more effectively develop products and services for high value creation. It removes the constraints imposed by relying excessively on internal resources.

Technology companies rarely act in this way, but it is expected they will do so much more in the near future. They will need to practise this approach occasionally, but regularly, in order to boost value-creation, alongside current, more internally focused innovation process.

Figure 6.1 illustrates how distributed innovation is put to work. The starting point is the market. On occasion, the firm must define how to shape the business by 'groundbreaking' or 'high impact' offerings – products and services for that market. Identifying and selecting these offerings mobilize focused efforts within the firm. The business case of the selected offering shows the highest potential for sustainable

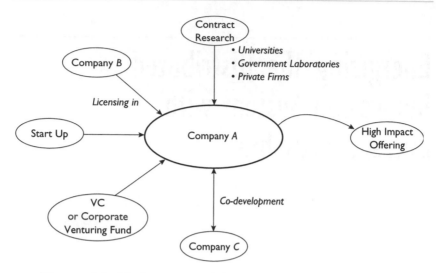

Figure 6.1 Market-oriented distributed innovation system.

value creation, while being consistent with the product strategy of the firm.

This approach is carried out from an entrepreneurial perspective: just as the entrepreneur 'sees' an opportunity in the market and then marshals resources to address this opportunity, the technology firm identifies selected 'high impact' opportunities.

The firm mobilizes the resources to develop the targeted offering; among these resources is a technical component. In distributed innovation, the technical resources come largely from the world of external technology. They flow to the firm from the components discussed in the previous chapter and shown in Figure 6.1. In this way, the firm mobilizes a much broader technological base and has more options available to carry out these developments. There are positive side effects as well, which will be discussed later.

In order to identify and select the offering to be pursued, a broad cross section of the firm's managers and employees are mobilized, who then have extensive discussions, in an iterative way. They are backed up by an excellent knowledge of the external environment: competitors, markets and technical developments. The overall process is driven by the CEO.

'High impact', high-value creating products have the potential of building a new growth business segment. This is certainly true, even if many new products do not rely on a breakthrough technology: the immensely successful 'Walkman' was not based on any radically new technical advance at all. The 'Swatch' is another such example. Arguably the same is true of the WiFi wireless radio connecting devices over a distance of about 100 metres. This capability will accelerate a tremendous proliferation of devices. Indeed, what is pursued are offerings presenting a great potential for sustainable, profitable business. Technology is there to provide a toolkit from which to select and use in developing successful products. In distributing innovation, it is breakthrough offerings that are sought not breakthrough technologies *per se*.

To a certain extent, Samsung Electronics followed this approach, although in this particular case the company did it not to develop a specific product, but to enter a whole new industry. Having set themselves that ambitious goal, Samsung engaged with various external technology partners, such as Micron. It bought licences, entered into alliances with firms and acquired companies in the electronics sector, while rapidly increasing its internal R&D activities. Through sheer tenacity in their commitment, the company became the world's pace setter in the DRAM (Direct Random Access Memory) chip business.

Boosting Value Creation by Innovating in a Distributed Way: Three Examples

Practising distributed innovation for enhancing value creation means the firm must allow the entrepreneurial spirit to flourish. Entrepreneurship energizes the development of solutions addressing targeted 'high-impact' opportunities. This approach calls on the CEO to be visibly involved in organizing and facilitating the process aimed at identifying and evaluating the most promising opportunity. All parts of the firm must be mobilized to define and prioritize opportunities in the marketplace. In a multinational company, a team of about 20–25 persons is appropriate, provided they are carefully chosen for their high skills and complementary competencies. The process of identifying and selecting the 'high impact' offerings is carried out with the help of tools, such as those discussed above. Let me take some examples to show how this may be played out.

In the first example, a mechatronics firm wants to participate in the future business represented by nanotechnology, which manipulates matter close to the atomic scale. The firm plans to use it for developing micro-devices for fibre optics applications. A first requirement is that the proposition must be made with enough conviction, so that the CEO takes personal 'ownership' of the issue. Intensive scanning of information and evaluation of the field are undertaken.

At the same time, the firm launches limited, but carefully focused, internal technical developments in order to secure a sufficient first-hand knowledge in the area, for the firm needs to be a knowledgeable buyer of technology. At some point, internal consultations, meetings and discussions, as well as external inputs, enable knowledge and information sharing; the issue is ripe for the CEO to launch a wide-ranging market-oriented initiative concerning innovative optical switching devices based on nanotechnology. An extensive series of discussions is organized, alternating between assessing the targeted business segment and the firm's capabilities, either existing or to be acquired. The search for a 'high potential reward' product is on.

The roadmap includes defining how best to secure the technical know-how required as quickly as possible. To this end, the sources of technology shown in Figure 6.1 are analyzed. Opportunities for in-licensing, partnerships with firms, governments or contract research laboratories, equity in start-ups, are identified and assessed. The pros and cons – access, quality and relevance of the technology, the patent situation, cultural fit between firm and source – of collaborating with these different actors are evaluated, as well as the preferred modes of collaboration. During this time, the activities of competitors are watched and monitored continuously, feeding input into the selection process.

All along, the CEO is clearly seen as being involved in orchestrating the process of defining the 'high impact' product and actively participating in all the steps, continuously fine-tuning and pushing for progress. This ongoing evaluation within the firm finally leads to the selection from a shortlist of one switching device. The decision to go ahead with the development of a business-oriented innovation is indeed risky, but if successful, will create a stronger, more differentiated position for the firm, which in turn will mean higher profit margins and enhanced value creation.

The development of the device then begins. It makes use of the internal capabilities, augmented by a considerable amount of external input from the selected external sources. The roadmap for the development specifies time-scales and milestones, as well as back-up plans to leverage the technology in other devices, should the ongoing development not deliver its promises. Very importantly, the project leader will have been prepared for the job by being heavily involved in the preparatory scanning, searching and in building the business case.

Another example would be fuel cells for automotive applications. Let us assume that car manufacturer X, observing the market and the regulatory environment, judges that the time is right for pioneering the intensive development of a hybrid car, to be equipped with a conventional gasoline engine, as well as an electric motor powered with fuel cells instead of batteries. The target year for introduction is 2009. Lessons from previous experiences with battery-powered automobiles have been learned. Following the State of California's zero-emission regulatory mandate, carmakers invested massively ($1 billion by Ford, for example) in the development of battery-powered electric vehicles. This turned out to be a complete failure due to the rigid requirements of the regulators and the limited range of the vehicles, as well as to the excessive leasing costs and an insufficiently developed infrastructure.

It now appears that the oil industry is moving to provide refueling stations, so that in due course the infrastructure problem will be solved. In order to pioneer this field and find an outstanding solution, massive development work will have to be done. The available capabilities inside company X alone will lead to a much too constrained choice of options, and as a result the effective way will be to act as the integrator of developments carried out by external contributors.

The first task for company X will be to build the business case and to define the desirable characteristics of the targeted product. This requires much searching and fact-finding in the market and the technology scene. Once the brief is defined for the product, the question will be: as partners in the distributed innovation system, who will provide the complementary units of technology? Is it the Ballard Power Systems company? This company is already 46 per cent owned by DaimlerChrysler and Ford. Another option is Hydrogenics. However, GM has already invested in it and their technology does not support the option that company X has retained. An alternative provider must be found. The technology watchers of company X identified Fuel Cells Inc. (an imaginary name for our present purpose), a young company in Israel which has a patented approach for an attractive option. After doing a due diligence on this company, company X negotiates an exclusive agreement with it and invests in collaboratively developing the technology.

Another piece of the technical puzzle is the metering system to feed the fuel into the cells. This is developed by a specialized technology firm under contract with company X, closely coordinating with Fuel Cells Inc. Finally, discussions with component manufacturers are started in order to get them to develop the kind of control systems that will be needed. A project team in company X coordinates the development of the various elements, articulating and synchronizing them with each other, as well as with internal developments. Distributed innovation demands a close coordination of the various components of the development.

In the small community of the world's carmakers, competitors will very soon learn that company X is on to something ambitious with fuel cell technology. But they will not know the exact intentions in terms of level of efforts and timing; the collaboration with Fuel Cells Inc. could be kept confidential long enough concerning critical details. In the case of the component supplier, it may well provide indications of X's intentions to third parties. This is a risk which must be taken, and it is not that great; innovation transfer does take time, and competitors will need time to switch technologies and catch up with X. In addition, company X has a secret weapon; in order to generate a new identity associated with fuel cells as a source of power, it decided that their new model will have a very distinct car design. It thus asked several studios to come up with a very new look for the car body. The novelty of the product will thus be branded by a unique car design.

A third example deals, not with a product, but with a 'macro' sector. Let us consider the entire healthcare sector in first world countries. In these countries, health costs are high and rising at a rate which is alarming to governments. They already represent 12–13 per cent of GNP in the USA, even with a relatively poor overall efficacy: life expectancy is lower in that country in comparison with other industrialized countries. There is a compelling need to come up with major, systemic innovations that would help provide quality healthcare at a lower cost.

The huge scope of this need for society would justify mobilizing the creativity and resources of many sectors. Among others, innovative approaches would include technical areas, such as prevention, diagnostics, therapeutic treatments and ICT – information and communication technologies. In this case, would it be helpful to have an entity taking the leadership to scout and bundle the appropriate technologies and know-how in an attempt to respond to this huge need? Who should this be: pharmaceutical companies, health insurance firms, governments, regional groupings such as the European Union or a national or international consortium of some of the above? Who would fund the efforts: health insurance companies, pharmaceutical companies, governments, international organizations? Such a need-driven initiative could create a powerful dynamic and emulate much innovative activity as a fascinating application of the concept of distributed innovation.

Going back to the 'micro' level of the firm; as indicated before, Samsung Electronics used this distributed innovation to bootstrap itself into the advanced electronics industry. As discussed, Generics is a model to emulate in leveraging commercializing technology by trading with external participants. Technology start-up companies draw extensively on external inputs to develop their products in the early phase of their life. At that stage, they concentrate on developing their offering. As a result, 40 to 60 per cent of their activity is on innovation development. For this they buy external services because it allows them to save time, as compared to hiring and training new staff. Later, they gradually internalize the activities in order to have more control, often to an exaggerated extent; control does not have to be synonymous with doing things internally. They should keep a balance in favour of 'importing' technical expertise, so as to have more options and learn more about the external scene.

What is proposed is that technology companies should continually strive to identify 'high impact' offerings, and then, after a due process of selection, launch appropriate innovation projects, alongside the other more 'conventional' developments essentially drawing on internal resources.

Innovating in a distributed way means an integrated, proactive process leveraging a substantial amount of external sources in order to develop carefully targeted 'high impact' innovations. Most companies still rely extensively on the internal innovation process, while occasionally making use of some of external inputs. The examples below of Intel, Nokia and of the pharmaceutical industry illustrate how certain companies heavily rely on external inputs, but in a piecemeal fashion. These companies are only separated from practising 'distributed innovation' by the crucial step of proactively bundling external and internal technologies in the 'seamless' way described above to pursue occasional groundbreaking innovation projects. They are therefore natural candidates for early adoption of this approach.

Intel: Innovation Inside?

The world's leader in semiconductor manufacturing, Intel, had a $26 billion business volume in 2001. Its headquarters are in the middle of Silicon Valley and almost all of its revenues come from microprocessors. It was founded in 1968 by former employees of Fairchild Semiconductor, Gordon Moore, Robert Noyce, Andy Grove and Leslie Vadasz. Intel encountered early success with its dynamic random access memory, DRAM, a field that Samsung later entered with remarkable success, as discussed in Chapter 5, and from which Intel eventually withdrew. Today, it may be said that Intel, together with Microsoft, drive the massive electronics industry, especially since IBM reduced its involvement with semiconductors in the 1990s.

Intel's founders were frustrated with the development activities of their previous employer, Fairchild. That company had a very strong research capability, but its corporate R&D was very much a case of the techno-centric 'ivory tower' of the 1960s, as described in Chapter 3. At Fairchild it took forever to go from new product idea to manufacturing. For the founders of Intel, the lesson was: let us concentrate on effective developments for the manufacture of high quality products.

The semiconductor industry depends heavily on technical breakthroughs to progress. This kind of research is an example of how blurred the boundary is between the categories of 'basic' or 'applied' research, as discussed in Chapter 1. It involves understanding matter at the atomic scale, but in a manufacturing engineering perspective. As a result, process development absorbs the majority of the close to $4 billion of investments in R&D made in 2001.

Particularly in recent years, Intel's approach to innovation has become increasingly open to outside inputs. Among the various channels shown in Figure 6.1, however, only two are used extensively: university research and corporate venturing. A third one is the consortium Sematech, mentioned in Chapter 5. Participation in this consortium appears to be more an opportunity for Intel to participate in a club to influence choices of the industry rather than to access external expertise.

Intel has developed a very close connection with university research. Each year the company invests $100 million in a number of 'academic' projects. From Dupont to now defunct Digital, many companies have this practice, usually housed in a 'university relations' department. Intel supports more than 200 research projects at various universities. It is more concerned than most companies however, in having a good return on these projects. The firm thus assigns an engineer to act as a close liaison person for each project. This enables the provision of guidance to the external project and also channels the results back into the company. This practice resonates with Procter & Gamble's campaign of 'connect and develop' for its product development activities of the 1990s. In this motto, 'connect' refers to connecting the firm with external technical developments.

The other channel, corporate venturing, was created in 1991 at Intel. It was a natural move for a company located in the middle of Silicon Valley, where the venture capital industry flourished. This unit is now called Intel Capital. In 2002, it had a portfolio of more than 500 investments, making it one of the largest venture capital funds in the world.[1] In this, Intel behaves like a venture capitalist (VC) and co-invests alongside other VC firms. Most of the investments are made in companies located in the USA, often in the Silicon Valley area. There are exceptions: one is the investment in Retro, described in Chapter 4. In this particular case, the investment is justified by the fact that the device relies on a sophisticated chip. Optical devices, in general, are also a promising area for the future.

Intel Capital primarily makes investments in companies that support the development of an environment that will enhance the usage of Intel's current products. Intel calls these 'ecosystems investments'. The criterion of good financial performance is, however, never far away. A small minority of the investments aim at securing windows on new technologies, according to the so-called 'strategic' rationale. In mid-2003, the valuation of the Intel portfolio is roughly $800 million.[2] This represents a precipitous drop from the $8 billion value of the stocks' heyday in the Spring of 2000. Such numbers greatly bias management's perspective on the fund.

Intel's microprocessor development is governed by Moore's law, mentioned in Chapter 5: the progress in performance of microprocessors along this S-curve is closely related to the ability to manipulate matter at a very small scale. At some point, this will mean going to the subatomic scale of quantum circuits. Another S-curve might possibly be biochips.

The approach of Intel to innovation has been consistent in having an internal focus on process improvements and a small set of external collaborations with universities. This is not the approach of distributed innovation, which relies on systematically managed, multi-actor projects for breakthrough developments. Indeed, the distributed innovation system would provide Intel with a way forward if it decides to launch a large development of one of the technologies mentioned above by bundling together external and internal forces in a major project. As indicated in Chapter 5, it is difficult for a company to shift from one technology to a radically different one, 'jumping' from one S-curve to another. Certainly Intel is a company that has the chance to be an exception to the rule. This is because few companies in this industry can match Intel's technical and management know-how, entrepreneurial energy, as well as its financial muscle and strong brand position.

Nokia

Chapter 5 described the remarkable business metamorphosis of Nokia over the years which resulted in a world-leading firm in the sector of telecommunications. Its overall business volume was Euro 30 billion in 2002 and has two main segments: mobile phones (more than 75 per cent of sales) and network infrastructure. As mentioned earlier, Nokia puts a very high emphasis on new product innovation and its internal development system is discussed below.

Like Intel, Nokia also makes use of several channels for accessing external technology. As discussed in Chapter 5, these channels are: in-licensing, collaborative developments, such as for GSM standard, Bluetooth and Symbian, as is customary in this type of industry. Another component worth looking into is corporate venturing. Corporate venturing at Nokia was set up in the late 1990s. The Nokia Venture Organization (NVO) was instituted in a unique fashion, in the sense that it is, in effect, a business development organization coordinating several distinct venturing activities. Importantly, the organization reports directly to the Nokia president, as suggested in Chapter 2.

The first group contributes to the advancement and growth of the existing businesses. It is, in effect, a strategic planning group. It looks at new opportunities, areas for business development and analyzes competitors. A second group deals with the growth of new business. It looks at ways of creating these businesses and how to link them to the existing ones. The third group is called the Early Stage Fund, which is an internal corporate venturing fund. It primarily invests in ideas coming from Nokia that have high growth potential. The fund money comes from Nokia alone. The fourth is an external venture fund which invests in areas known to Nokia, so that due diligence of the investments considered is facilitated. The primary objective is to have a high return on investment.

Nokia's internal innovation system relies on a yearly investment of Euro 3 billion in R&D. This represents 10 per cent of the company's sales volume. The R&D system includes 70 R&D units across the globe. Nokia's top R&D executive, Yrjö Neuvo, who encourages these units to be autonomous and audacious, says: 'people should not shrink from making mistakes'.

From this spirit of daring came many success stories, such as the Navikey user interface. This was a crucial step in making mobile phones more user-friendly by combining three separate buttons into one single bar. Matti Alahutta, president of the mobile phones division, did his PhD thesis in management on the challenges of growth in technology companies. He states: 'we allow teams to have their own space. People have to feel that they can make a difference. We try to encourage a small company soul in a large firm and we innovate all the time.'[3]

In brief, Nokia mobile phones has succeeded in creating a number of small, 'high energy' development units. It thus coordinates a set of teams internally distributed within the company as well as managing a fair amount of external collaboration. It is therefore in a good position to be able to practise distributed innovation, if it wishes to do so.

The Pharmaceutical Sector

Pharmaceutical companies have been pioneers in the extent to which they leverage external resources. They have a long history of reaching outside to complement their innovation pipeline. As indicated in Chapter 3, the drug development process is long and costly and may also be terminated abruptly by the discovery of an unexpected side effect in clinical trials. Companies therefore tend to hedge their bets by tapping external sources of new molecules and innovative approaches.

Johnson & Johnson (J&J) is an example of a company which has been successful in the difficult process of internalizing external innovations. It has been able to acquire a number of external ideas and to develop them into its own products; examples include disposable contact lenses and glucose monitoring. This demonstrates a low 'not invented here' barrier to external inputs. Consistent with this is the fact that J&J is managed as a decentralized, agile organization.

Pharmaceuticals have also relied heavily on the external world of science and technology in order to acquire enabling technologies. Genetic engineering is such an example. Several years ago, large pharmaceutical companies purchased equity in companies working in the area of gene therapies. Roche bought California-based Genentech, while Novartis (Ciba at the time) invested in Chiron.

Pharmaceutical companies continue this momentum. Roche has many agreements in place with genomics companies; the agreement with deCode, for example, covers twelve diseases. Like Intel Capital, pharmaceutical companies continue to invest in start-ups: Novartis invested in 18 start-ups in 2001. The same year, Danish Novo Nordisk invested in nine start-up companies. Pharmaceutical firms may also invest indirectly by securing participations in funds, which themselves invest in start-ups. Either way, they secure windows on technologies embodied in the start-ups, which might be important to build future businesses.

In spite of the precipitous drop in the investing activity of the venture capitalists' industry in 2001–2002, investments in life-sciences start-ups, although also reduced, remained relatively strong. This was due to the fact that such firms generally rely on more solid, scientific innovations than the Internet-based businesses. It is also due to the fact that, if they are successful, these start-ups have a market, constituted by the pharmaceutical companies. First, the latter support the start-ups by entrusting them with research projects. Second, they have the cash and interest in possibly buying them in a trade sale at some point in time. Is the life-sciences area the next bubble? If so, it will probably be not as extreme as the previous internet-based one. This is because the industry relies on science-based, patented innovation and undergoes more 'robust' due diligence. Another key reason is that the corresponding development times of several years are much longer than just one year, which used to elapse between business plan and IPO in the internet business in the feverish 1990s.

In addition to connecting with start-ups, pharmaceutical companies make extensive use of external collaborations, which may involve in-licensing. The driving force for this is for a company to increase its portfolio of drugs marketed and sold by very expensive distribution channels: the pharmaceutical industry spends at least as much on promotion and sales of drugs as on bringing their own molecules to market.

External collaborations also include contracting out development to external partners, small firms, universities, or CROs – contract research organizations. In the latter case, this is primarily done for clinical studies, which represent an increasing part of the overall cost of bringing new drugs to market. It is anticipated that the pharmaceutical industry will increasingly outsource development projects and technical services. The primary reasons for this evolution are:

1.   flexibility, and
2.   reduction of capital on the balance sheet.

Certain forecasts estimate that an enormous 40 per cent of all pharmaceutical R&D will be outsourced by 2010.[4] This represents a volume of[5] approximately $13 billion per year. The number was $9 billion in 2001.

These numbers reinforce the earlier point, according to which start-ups are eager to develop a business of sophisticated technical services to be sold to the pharmaceutical sector. The services include research activities such as discovery of active molecules, and services such as pre-clinical and bio-analytical testing, informatics or high throughput screening.

Of all industries, the pharmaceutical sector has been the one that has had the largest practice of outsourcing R&D. It is, however, not practising distributed innovation in the sense we have been describing, that is, a project aimed to solve one single targeted product by combining diverse, external and internal technical capabilities. Instead, the pharmaceutical approach has been to *successively* leverage several channels of technologies in the course of the 10–12 years development cycle for a given drug. This may involve leveraging ideas from a university professor, contracting research work in the early stages, then years later, clinical studies. This does not present at all the same challenges as managing one single project simultaneously engaging various sources of technical knowledge.

Pharmaceutical companies often work in parallel on several alternative routes for the treatment of a specific condition, such as hypertension, in the cardiovascular therapeutic area. In this case, there are really several projects with different project leaders running simultaneously, instead of one single, integrated effort. This might involve an in-house project following a given therapeutic route, while an external approach would follow another. This is similar to what automobile makers do with car design: they may have an internal design department working on one model, while at the same time an external design studio is asked to provide solutions for the same model. The results of these duplicated efforts are compared and the best outcome is chosen, which may well incorporate elements from the other approach.

Practising Distributed Innovation

In brief, these three examples present complementary elements: Nokia operates something like a distributed innovation system *inside* the company. At the same time, Nokia, like Intel, substantially engages with external partners along certain specific channels, that is, co-developments and corporate venturing. On the other hand, pharmaceutical companies reach out extensively for external technologies, playing with the complete keyboard of channels available in the distributed innovation system. In short, Nokia has the distributed dimension, while pharmaceutical companies have the multi-channel dimension.

Because they already master elements of that system, these companies are in a good position to function within this system by managing innovation projects that would draw on external, as well as internal, technical capabilities. It is likely that the pharmaceutical companies will practise distributed innovation extensively. This is due to their existing broad experience with leveraging a multiplicity of channels, as well as the constant and compelling need to try to enrich their drug development pipeline. It is therefore recommended that these companies bundle together external and internal technical expertise to develop pre-defined drug targets.

Thus innovating using a distributed approach involves the following steps:

1. *Identification.* Identify the 'high impact' offerings or products for the firm. Build a business case for each of them.

2. *Selection.* Select the most promising offering appropriate to the firm.

3. *Technology mobilization.* Seek out external technology for developing the selected product. Technical expertise is found in external sources, such as other firms or start-ups, universities or CROs – contract research organizations.

4. *Development.* Develop the product by combining internal and external *contributions*.

5. *Production and distribution.* Manufacture, market and sell the product.

Such an approach is carried out alongside innovation projects internal to the firm. It is activated when the company needs to create a groundbreaking opportunity to shape and grow its future business. Distributed innovation constitutes a new way to envisage innovation. It therefore creates new demands on the firm practising it. These demands are diverse. Having such a complex 'innovation supply chain' raises issues for each of the steps given above. The implications of these are discussed below.

The first requirement is a sophisticated system for gathering techno/ business intelligence. Effective scanning of the external environment must be an ongoing process. This activity is carried out by companies today as a way of watching market evolutions, competitors and technical threats to the firm's business. In distributed innovation, this watching must also concern possible sources of technology. It means that the technical function is fully involved and constantly *aware* of the opportunities for sourcing specific external inputs.

The objective is not to monitor a very large number of information sources, which is too time-consuming and, not effective. Endless working days could be spent on the Internet, checking data banks and company sites. The first step is to identify what the *key sources* are, and, importantly, those providing the most reliable, exhaustive, up-to-date and sophisticated *intelligence*. This process of selection of sources is performed by consultations, discussions and peer advice, and it must be rigorous in order to produce a shortlist of reliable, informative sources. The objective is also continuously to evaluate these sources, while watching out for possible new sources emerging.

As in any intelligence gathering, information sources include colleagues, individual experts and advisors, patent searches, newsletters, trade shows (Canton Fair, CeBit) and conferences, publications, data banks and Internet sites. On the one hand, these sources are monitored to identify market trends on which to base evaluations internal to the firm to define the targets to be pursued. On the other hand, they are monitored to identify those repositories of technical knowledge that may eventually become the building blocks of the distributed innovation system for the company.

The scanning activity is carried out by a well-coordinated group of technical and business staff. This 'radar' must be high performing in two directions: first, it needs to detect market intelligence to provide the best possible judgement for the choice of new offerings, and second, it must be able to identify effectively the sources of technology complementing the firm's own technical pool. The group must have strong leadership and provide frequent contacts with the innovation board steering the total innovation development process, as will be discussed below.

The second requirement is for the firm to have an effective process to identify, assess and select the most promising products. This requirement is important even when innovation is developed internal to the firm. In the course of this selection process, management tools and processes are utilized. In the new situation of distributed innovation, they must be deployed with the keen awareness of two additional considerations. First, the search is for 'high impact' products, which have the potential of creating high value and a stronger competitive position. Second, this is done with the knowledge of the technical options available outside the company, as these contribute to the shaping of the definition of these products.

The third requirement concerns the coordination of the development of the product. Integrating external and internal contributions involves a much more complex process than managing a predominantly internal process. We return again to the crucial importance of the choice and the support of the project leader for the success of the development. Of particular importance is the ability of the project leader to manage the complexity of the development process, effectively grafting the external inputs onto internal developments so as to grow a viable tree as quickly as possible.

Support for such projects and their leaders must be provided by the CEO. In the high risk/high rewards endeavours we are talking about, the CEO must be clearly seen to be resolving the innovation paradox by having a personal stake in the success of the new projects. As will be discussed later, this implies an appropriate reward system.

In addition, the innovation board, as described in Chapter 3, contributes during the negotiation phase with external providers of technology, whether for licensing, collaboration contracts or equity investments. The main role of the board remains to act as a 'control tower' guiding, evaluating and, if need be, interrupting the course of the development projects.

A prerequisite is the *willingness to buy technology from outside*. In this context, the debate on 'make or buy' depending upon whether the internal pool of technology is 'core' is somewhat theoretical. The notion of 'core' technology is perceived differently by different companies: while Daimler-Benz out-sources fair amounts of its engine development, other carmakers jealously keep this activity within the firm. The whole idea of concentrating on 'core' activities is probably yet another manifestation of 'knowledge inertia', as will be discussed below. The central question is: if an attractive opportunity is identified, and the best way to develop it rapidly is to buy outside technology, then the firm should go ahead and buy it, if the price is right. Rather than 'make or buy', the slogan should be 'buy and learn'. This brings us to the third prerequisite.

Effective knowledge management is another prerequisite. Far from being immune to external inputs, the firm must have a strong capacity to absorb knowledge coming from outside. The outward reach for markets and technology must also extend to learning and internalizing knowledge. In this sense, the right attitude is that of certain Japanese companies, which go into alliances with other firms with the clear objective of learning from their partners as much and as fast as possible.

The high risk/high rewards product developments through distributed innovation indeed mean large investments. They also mean longer time horizons: three, four, five years or possibly more. In our stop-and-go corporate world, how do we sustain commitment over such long periods of time? This will be facilitated if the various stakeholders in the firm 'swing the pendulum' away from 'short termism', as discussed in Chapter 2. Within the firm, the process of development and its rewards must be aligned. For the CEO and for the team managing a project, the bonus system should be governed by the progress of the project. The bonus should thus be given in a staggered way, governed by the successful reaching of a sequence of checkpoints set for the project. The remuneration committee of the company board should administer these rewards, since they involve the CEO and mean potentially critically important developments for the future of the company.

The other implication concerns managing the human factor of technical professionals. In the distributed innovation concept, R&D staff constitute the 'technology brain' of the firm and, in particular, do the scanning and scouting of external technologies. They will also, carry out an 'audit', or 'due diligence', on these technologies and make a recommendation as to acquiring them or not, using the various channels discussed. This is potentially a totally schizophrenic predicament. On the one hand, as technical knowledge workers they see their competence in a given specialty area and are tempted to continue working in that specialized area. Yet on the one hand these same persons might now have to recommend to their employer buying a technology that would make them obsolete! Chapter 7 will discuss the human factor issues further.

The approach of betting on distributing innovation development by drawing on external technologies is expensive in terms of capital and human talent. It also entails substantial risks, although they do not go as far as 'bet the company' risks. This is the price necessary to pay for obtaining enhanced access to a wide range of technical alternatives, articulated with and complementing internal developments. What is sought is not time efficiency, but *effectiveness* in the innovation and value creation processes.

Should the venture fail, the experience of the development will, at the very least, bring useful learning to the staff concerned. It will effectively contribute to further opening up the company to the outside world. For example, technical development activities will have provided an opportunity for the company to calibrate its activities with those of external actors.

In addition to the business benefits, successful ventures will also enhance the entrepreneurial spirit within the whole company, diffusing boldness and agility throughout the firm. The positive impact will affect the overall perception of the firm. As successful initiatives are launched over time, these characteristics will be reinforced and the firm will become more attractive to precisely the kind of entrepreneurial talent that it needs but which is in high demand. By showing a more entrepreneurial attitude, the firm will be able to recruit the talent it is ardently seeking. In this way, thanks to the venture-creating component of its activity, the Generics company has been able to attract good scientists with a keen business sense and high entrepreneurial spirit. Part of this attraction is the possibility of becoming involved in a project which may one day be turned into a venture and, eventually, a company. The same rationale and substantial benefits hold for firms like British Telecom with 'Brightstar', discussed in Chapter 5. This underscores the importance of the profiles and levels of motivation of the personnel involved in the type of ventures typical of the distributed innovation system. This human factor will be discussed in the following chapter.

Conclusion

In brief, technology companies are expected to rely on distributed innovation in order to develop offerings able to catapult them on the higher grounds of competitiveness. For these occasional, but carefully prepared ventures, the firm acts as an *integrator* utilizing external and internal technologies 'seamlessly'. This five-step approach improves the effectiveness of the innovation process by removing the constraints that result from drawing mainly on resources internal to the firm.

This approach is also a powerful way to impart an outward orientation and an entrepreneurial energy into the firm. These constitute handsome side-benefits that are crucially important elements for firms if they are to compete effectively in tomorrow's world.

Notes

1. www.intel.com/capital.
2. *Ibid.*
3. Private communication, 4 November 2002.
4. *Pharmaceutical R&D Outsourcing* (Reuters, 2002).
5. *Ibid.*

CHAPTER 7

The Crucial Human Factor

The critically important quality of output of innovation projects is a direct function of the talent and motivation of the staff who carry them out. A motivated researcher not only does research, but also *finds* new approaches and solutions. Without motivation, the innovation engine burns investment money without producing any useful output.

If the management processes described for the distributed innovation system in the previous chapters are to work at all effectively, the staff concerned absolutely must have the energy and excitement which provide high levels of motivation. For motivation to flourish, a basic requirement for management is to be attentive to knowledge workers. This is particularly true in the first months of a new hire on the job. Intuitively and by trial and error, sensible managers develop their own common sense practices to achieve this attention.

A second motivator is the management style. Like artists, staff dealing with the uncertainties of innovation projects must have a supporting and empathic management. Accordingly, management should act as a coach and practise walk-around management. One aim of the coaching activity is to develop business sense and entrepreneurial spirit in the staff. First-line managers have a particularly important role in this development.

Finally, a strong promoting agent of innovativeness is a rich diversity among the staff and organizations involved in any given development project. Here also, management must know how best to leverage the richness of diversity.

Be Demanding and Supportive

The first weeks on the job influence the motivation of the new hire for a long time. First-line managers must be particularly attentive during this period. This is illustrated by the following fable.

After obtaining her doctorate in electrical engineering, Dr Joanne Talent vacationed in Cambodia and Laos, and then started her first job. On a glorious September morning, she arrives at the headquarters of Computech Corp. for her first day in the office. The company is located near Saint Petersburg, in Florida. It is a sunny, balmy day. The building has an impressive glass and steel architecture.

No, the receptionist has not been notified of her arrival, but her name is on the roster of employees, so she is allowed to enter. Her boss, Al, from the Research and Development (R&D) Department is not available. He is at a weekly staff meeting for another hour and his assistant is not sure where Joanne's workplace is.

Joanne finally finds Stéfane at the photocopying machine, who had participated in one of her hiring interviews. He had forgotten that this was her first day on the job. Where is Joanne's desk? Stéfane points to one which he thinks has been allocated to her, but there is no sign to confirm this and her name is not on the office door. 'Why don't you come and have a cup of coffee?' says Stéfane. This was the first ray of light in a disastrously depressing welcome.

Coming back from his routine meeting, Al asked: 'What's the big deal?' of his concerned assistant, who had tried to convince him to leave the meeting and come out to welcome Joanne. Failing to make the small effort of decently organizing the arrival of a new hire, when the company's annual report boasts that 'the most important asset of the company is our people' will have a lasting effect on the psychic energy flowing between employee and employer. Al should be moved to a staff position, because his behaviour demonstrates that he does not have a clue about the basics required to inspire a modicum of motivation in his contacts with people.

Computech had invested considerable time and energy screening and selecting candidates before offering the job to Joanne. How did the management then fail in the common sense of organizing her arrival to make her feel welcome? Variations on the theme of this glum episode are very frequent: firms agonize in endless meetings when deciding a 100,000 Euro investment in a piece of equipment, but err in the most important process of all: selecting, hiring and integrating new professionals.

At the other extreme is the practice of large Japanese companies, which celebrate the arrival of new hires at the beginning of each fiscal year with a solemn welcome by the CEO, then extensive introductory sessions on the company's history, tradition and activities, followed by spending time in various parts of the firm. Appropriately, the Japanese language has no word for 'employee', the equivalent word is 'shaen', which means 'member'. This attitude may be considered excessively paternalistic, but it marks very positively an important step in the young professional's life and contributes powerfully to creating commitment and loyalty in new staff. True, this practice is to be seen in the context of lifetime employment, which, although less absolute than twenty years ago, is still largely the rule in most large Japanese corporations.

Integrating New Hires

'The worst is not always sure ...' In contrast with the Computech episode, Laura, one of the seven unit managers in the research organization MatLab, was careful to organize a party soon after a new hire had joined her unit. This get-together at the workplace provided an excellent opportunity to introduce the new staff member to the community and to explain the rationale for his or her hire. It also sent a very positive message of welcome, celebrating a new arrival to strengthen the team. It is better to give parties when people join than when they leave the company, Laura thought. All her managers–colleagues agreed that this welcoming party was a great idea, congratulated her on it, but nobody followed Laura's good example. Privately, she reflected on how human beings do not always emulate what they know to be a good example, far from it.

The same manager also had the habit of having a new hire spend the first three months on the job sharing her own office. In this way, Laura was readily available for questions and she could gradually inform the new hires and introduce them to people coming to her office. The hires concerned said afterwards that this was extremely useful and effective in accelerating their integration into the team. Here again, top management as well as her peers congratulated Laura on an excellent idea, but nobody imitated her practice.

Common sense fully supports Laura's efforts to welcome and integrate a new employee. The reality is something else: a new staff member arriving into a unit definitely 'detracts' management from the ongoing stream of daily activities. It represents additional work for integrating and training. Management's temptation is mostly to exile the newcomer into sharing a remote office with an almost equally junior colleague, hired six months earlier, so that the one-eyed leads the blind.

One particular temptation should be resisted – the low-risk option of delegating an unchallenging task to new hires, just because this will require less supervision and management attention. Newcomers should be assigned clearly articulated tasks, do-able but challenging, as well as requiring intense interaction with colleagues. The place of these tasks within the larger project must also be clearly explained, so that the big picture is understood by the new hire. Do not succumb to the temptation of sending a newly arrived Joanne into isolation, burying herself in the library to do a literature search. Instead, involve her in a new challenging project that will cause her to make mistakes, from which she will learn, while closely interacting with her colleagues.

The Blues of the New Hire

If the first days on the job are critical for the new hires, another crucial period is four to six months into the job. The excitement, hopes and expectations of the first days gradually give way to frustration in the fresh PhD. Things are not turning out quite the way he or she expected, and many aspects of the job have nothing to do with focused, academic work. The new environment does not value scientific competence *per se*. Al's statement: 'We are here to produce great quality chips, not publish papers' may sound harsh to a newly minted PhD. Instead, scientific expertise is one of the many competencies required: ability to work well with team colleagues and to practise 'business speak', presentation skills, for example. Scientific publications represent a low priority and new hires have to become aware that certain information is sensitive. Confidentiality and the mechanics of patenting are a business approach to protect a commercial position. For lack of proper warning by peers and management, young R&D professionals occasionally blunder by prematurely divulging facts at a meeting or in a conference, thus destroying any possible patent position for the firm.

New hires soon find management processes frustrating, in that they take much time and energy at the expense of technical contents. Management indeed sounds like a 'dirty word', an overhead activity. Peer recognition, a cornerstone of the scientific community, is not so central any more. The chemist, the physicist, for whom the scientific discipline constitutes a large part of their identity, feels somewhat of an orphan. The relatively lonely work in academia is succeeded by the need to interact with many different people with various profiles in marketing, law, finances. Gradually, a place must be found among the individuals who make up the unknown world of the firm. The highly complex transition from academic pursuits to business-oriented innovation in a firm's environment must be carefully monitored and accompanied by management. Otherwise this transition can result in a drop in the motivation level, as illustrated below.

Conscious of this pattern, the direct manager of the new hire anticipates this period of 'blues'. A timely and trusting dialogue allows the new employee to let off steam and to realize that such a predicament is common, a fact which comes as a somewhat reassuring discovery. This close communication helps usher in a 'recovery'

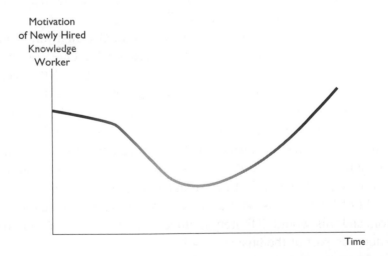

Figure 7.1 Motivation level of the newly hired knowledge professional in the first months on the job.

phase, allowing the new hire gradually to come to terms with the actual situation. Not having this 'six months later' conversation will cause motivation to further deteriorate and the recovery will take longer and be shakier.

The trajectory in Figure 3.4 illustrates the evolution of technical professionals with time, as well as that of the R&D function in recent years. A young PhD, fresh from university, enters the firm through the top corner of the triangle. The R&D function acts as a 'buffer' between the 'academic' world of science and technology, and that of business. With time, the professional gradually becomes less encapsulated in the sphere of scientific excellence and challenging technical contents. He or she develops more of a business perspective, helped by the coaching and mentoring activity of first-line managers. Such a professional thus chooses the managerial track and is in due course likely to leave the R&D function for other positions such as technical sales or manufacturing.

On the other hand, a minority of professionals elect to continue to be primarily technical contributors, remaining in the vicinity of the top corner of the triangle in Figure 3.4. In order to reward them in a way that is equivalent to the managerial path some companies – Intel, ICI and IBM for example – have in the 1970s instituted the so-called technical ladder. The latter includes hierarchical steps – scientist, senior scientist – paralleling those in the managerial ladder. At present, such dual-ladder systems are less fashionable, because they imply notions of hierarchy and stability.

What can be done to prepare this radical transition from the curiosity-driven to the business-motivated R&D? One answer is to make science and technology students more aware of the world of enterprise during both the final undergraduate year and at graduate level. This need not be time consuming – one or two hours per week for six months or so. If carefully prepared and coordinated with courses in the 'hard sciences', such an introduction to the activities of the commercial firm would be a useful way of planting a seed for the future. The objective is to acquaint engineering students with the world of business, as well as with commercially motivated innovation, and this would definitely include teaching the nature and role of patents as part of the business toolkit.

Later in this chapter we will discuss how best to accompany knowledge professionals in their transition from science and technology to business. First, however, it is important to consider the management style most appropriate to managing innovation. As an early example, here is an eighteenth-century quote on the duties of a plant manager. It is an extract of the 'Rules for the Royal Manufacture of Saint Gobain':

> He shall devote all his ability and application to manufacture good glazing and avoid defects which are but too frequent. He shall listen to all ideas on that matter whoever they are coming from. He shall make mature reflections and take the benefit of it, if he finds them good. He shall beware of falling into the mistake of some of his predecessors who, by fantasy and presumption, imagined that all which did not come from them could not be good.

This text is dated 10 December 1728. It is remarkable that almost three centuries ago, this statement singles out some of the concerns of today's management, such as quality and low reject rate, the NIH – Not-Invented-Here – syndrome, the necessity to have an open mind and the willingness to listen to suggestions.

What Management Style for Managing Technical Professionals?

In order to thrive in their work, technical knowledge workers need the supporting style of 'management by walking around'. This is best suited to maintain an ongoing dialogue, while helping staff to develop more of a business sense and an entrepreneurial perspective. This should be done in a coaching manner.

The Walk-around Manager

Because of the importance of the scientific component of their work, R&D professionals need frequent exchanges – not only with their peers – of ideas about technical matters and the latest publications or conferences in their field. Because of the uncertainty central to their work, as discussed in Chapter 3, they also need contact with, as well as support from, their first-line manager. An open door policy and a 'walking around' style of management are best suited to managing knowledge workers.

This must be done in a climate of openness and trust, while at the same time demanding commitment and quality of output. The manager should constantly and informally seek information on the advancement of projects, their progress and difficulties, in order to be able to provide guidance, contribute timely technical input as well as appropriate business intelligence. These informal contacts can take place in the laboratory, in the offices, or in neutral spaces such as hallways, the library, the cafeteria or by the photocopying machine, occasions for serendipitous meetings similar to the agora, the forum or the village well in earlier times.

The manager must show empathy towards the 'masters of the craft', while understanding the substance of their project work. For that reason, it is rare to see non-technically trained persons remain long as managers of technical development units. It is, however, advisable for a manager to leave a technical development unit in order to go to a business job and to come back later to the unit. Such a job rotation can be extremely powerful in bringing business sense to the technical units. Japanese technology companies such as Canon, Hitachi and Sony use this shuttle practice as part of their usual wide job rotation between functions, a by-product of the life-employment system in Japan's large companies. Developing staff by such job rotation is a worthwhile investment in the firm's employees, 'shaens', who will be with the company for a long time. The Brussels-based chemical firm Solvay also practises this shuttling of managers by having certain project leaders stay with the development and scale-up of a chemical engineering process, all the way to its industrial application. Such managers often go back to the laboratory to lead another project.

A unit manager supervising 35 staff carrying out some 60 different projects thus spends a minimum of one hour every day keeping his 'ear to the ground'. This may be perceived by the busy manager as a high investment of time, but it will be richly rewarded, because:

1. Gaps and mistakes in projects are most likely to be caught early, thus saving a lot of time and energy later; managers with good judgment can best improve productivity of their teams by *anticipating* possible problems and acting on them in an effective way.

2. A rich interchange is part of the supportive, motivating and trusting coaching role, suitable to managing knowledge workers.

The Manager as Coach

The first-line manager is not the all-knowledgeable expert, but may strongly contribute to enrich the researcher's environment. He or she really acts as a coach, available when needed, discreet when the researcher needs to concentrate on a delicate phase of the experiment under way, engaging in dialogue when the person is ready to receive input. One of the most counterproductive attitudes for such a manager is to display a patronizing attitude, with statements such as: 'I give my staff pencils, computers and the equipment they need and I let them play in their own way.'

In these times of a highly turbulent business environment, managers are so busy with operational tasks and staff meetings that they find it difficult to *find time* for nurturing a trusting and inspiring relationship with the staff working in their units. 'Being busy' is a convenient excuse for allowing the urgent to take precedence over the important; it is crucial that abundant interaction takes place between the manager and the staff in a coaching mode, in order to accelerate both personal development and, in particular, business sense and entrepreneurial drive. Making this a priority will make it possible to develop a number of technical contributors into the role of shaping the business-creation process.

On the other hand, it would make sense to have younger managers act as advisors to senior management, in order to convey the young generation's perspective on both the firm and the markets. This is particularly helpful in sectors such as fast moving consumer electronics goods and entertainment, where the younger generation has a major impact on the market. The remarkable success of cellular phones for i-mode electronic messaging with Japan's young generation is a testimony of how important it can be to understand the life-style of teenagers.

Coaching can come from outside the firm. External 'mentors' can work with a small number of individuals in total confidence on all topics of interest concerning their professional and personal life. In this process, the 'mentors' also form their own perspective of the main issues facing the firm and thus offer valuable feedback and guidance to management. An increasing number of companies, in Scandinavia and Great Britain in particular, are calling on the services of such external coaches and mentors.

An example of the value of 'mentoring' is provided by one of the units of the Swedish telecommunications leader Ericsson. Since 1996, the unit has been using this approach in collaboration with IMD – International Institute for Management Development – in Switzerland. Mentors were carefully selected, both from Ericsson's senior managers and non-Ericsson managers, and came from very different industries, from fast-moving consumer goods to branded products. The mentoring activity forms an important element of a comprehensive management development programme. It goes on for well over a year, involving face-to-face meetings with the mentor, as well as contacts with electronic communications. Initially, these take place predominantly at the request of the mentored manager and with a mutually agreed frequency. The confidential conversations constitute a real give-and-take, as the learning goes both ways. In fact, it is a prerequisite for a successful mentoring relationship that the mentor sees great value in interacting with the younger managers.

Almost a hundred Ericsson managers have benefited from this programme, which allowed the company to 'turbo-charge' the individual development of managers. This is especially critical for firms in fast-paced industries, where young managers are given early promotions in times of rapid growth.

What is Needed is More Researchers–Entrepreneurs!

A rare breed of researchers combine technical know-how with a vision of how they want to change the world. Their pursuit is thus truly about change and challenging the status quo. The dual capacity for understanding technology and business was found in the archetype of role models, Thomas Edison, in his Menlo Park innovation machine. But think also of the nineteenth century Pasteur, fighting the medical profession to convince them of the role of germs, while at the same time acting as advisor to the beer industry in a contribution critical to that industry. In order to bridge the gap between the technical and business worlds in technology firms, some of their knowledge workers must be *researchers–entrepreneurs*. According to conventional wisdom, these two profiles do not naturally coincide: the perception is that the researcher is concentrated on specialized technical aspects, while the entrepreneur is a doer oriented towards the marketplace.

What is an entrepreneur? Entrepreneurs see an opportunity in the market and marshal resources (people, know-how, capital and equipment) to address that opportunity. An entrepreneur has a passion for his or her project. At times, this drive may be blinding so that attentive guidance is required, while a positive attitude of support is welcome, as the entrepreneur's quest is lonely. The entourage of the entrepreneur channels and directs the entrepreneurial energy, but above all it contributes to keeping this energy alive: without it, the momentum vanishes.

Entrepreneurs are opportunistic in progressing their ventures. Good entrepreneurs manage risks astutely. They make the best out of limited resources and have a Spartan professional life: the frills and the trappings of power for glamorous managers are not for them. Their company's headquarters are modest buildings in a low-cost location.

Researchers–entrepreneurs are eager to see their ideas impact on the world and wish to take their idea themselves all the way to the market. All the same, they are often happy when another person, an entrepreneur, does it for them. Regions such as Silicon Valley and Cambridge, UK, are exemplary in the way they act as magnets attracting a population with this profile. In Chapter 4, we saw that an important side-benefit of Generics' business model is precisely to have this attraction. Within firms, the population of researchers represents a group of people, in which an entrepreneurial spirit can flourish. To encourage this entrepreneurial spirit the environment in the firm must fully support it.

It is remarkable that inventors often have a vision from the outset on how an industry can be developed to turn their inventions into commercial enterprises. Soon after inventing the moving pictures, in 1895 Louis Lumière organized the first projection of a film at a café in Paris. It was a paying performance and the success was immense. He had not only invented the cinema but also a whole new entertainment industry, the leadership of which eventually would migrate to the USA. He had a clear idea of how to combine the various links of the production chain, such as actors, filmmaking, production, distribution. Unlike Thomas Edison, however, Louis Lumière was not a role model for entrepreneurship, for he was uninterested in turning his inventions into a business. Instead, he went back to his laboratory in Lyon to continue inventing other processes and artifacts, including the loudspeaker.

As will be seen later on, developing the business sense of technical professionals and taking some of them down the path of entrepreneurship must be a top priority for management. This perspective begins at the hiring stage. As in all human organizations, managers should consciously hire staff better than themselves – but how many actually do it? Furthermore, a number of candidates with an entrepreneurial profile must be attracted to the firm. The breed of researchers–entrepreneurs is indeed what makes the success of regions such as Silicon Valley or Cambridge, UK. It is most likely also to be found in technical service organizations that propose innovative solutions to client firms with a view to improve their competitiveness. Such organizations include Battelle, Fraunhofer and Generics. Quite a challenge for the innovator to mobilize support for his or her idea, on the strength of business arguments supported by a mere patent application or a proposed approach! The client who proceeds with funding an innovation project takes a leap of faith, based on the perceived potential benefits and the credibility of the project team: the latter is selling a risk, a much bigger challenge than selling an insurance policy.

By being constantly exposed to private industry, these technical professionals develop a market-oriented mindset; they understand fully that they do not carry out projects only for making a technical achievement, but for their client's competitiveness. The first thing to do is to define the problem at hand together with the client, since the latter has often not framed his own problem correctly. Various approaches are explored and discussed. Only then can work on the most promising approach to the solution begin.

A large pool of know-how is accumulated within the technical service companies. Innovativeness means also matching parts of existing technology in order to find a way to solve the problem at hand. This internal technology transfer makes the know-how from different technologies flow towards a solution, when the organization is truly transdisciplinary and multi-market. To illustrate such a technology transfer, consider the example of a patented process to coat metal sheets. The process was first proposed to steel makers as a new low-cost way of applying a corrosion protecting zinc coating to steel sheets, to be ultimately used for automotive bodies. The feasibility was successfully proved. The new process was attractive because it was fast, flexible and provided high quality coatings. However, the scale-up to larger widths of steel sheets would require extensive developments. A more promising application was then identified in the course of contacts with the electronics industry; the process in question offered an attractive solution to apply metallic coatings on strips that would be used in the manufacture of lead frames (substrates for microchips), and this development was pursued for a large Japanese company. Following the principle of flow of technology transfer, the final application had now shifted away considerably from the initial ideas.

This ability to *see* the business benefits of technical knowledge is central to the techno-entrepreneur. It is this ability that must be developed, as will be discussed below.

First-line Managers Must Effectively Develop an Entrepreneurial Business Sense

Developing the entrepreneurial spirit in R&D professionals is a multi-level process. First, there must be continuous effort to remove bureaucratic barriers and inflexible ways of doing things, which discourage entrepreneurial energy. Second, appropriate role models and company history, constituting the 'folklore' of the firm, support the spirit of entrepreneurship and are also powerful incentives. Examples include the image of the garage, where the founders of Hewlett Packard first worked, Battelle's pioneering development work which led to the creation of Xerox (Haloid Corporation at the time), 3M's invention and development of the 'post it' sticky paper, that does not stick permanently.

A third level is the leadership of first-line managers in fostering an entrepreneurial outlook in their units. Encouraging their R&D professionals to adopt an entrepreneurial mindset creates a momentum that may well encourage a strong taste for independence in the staff. Some of them may thus elect to leave the company in order to apply their newly discovered vocation by starting their own business: they may want to create their own jobs instead of remaining in the one they have. Like any change process, encouraging entrepreneurship thus opens a Pandora's box. While regretting the loss of energetic talent, the firm should be positive about this and make all efforts to remain in healthy contact with the departing entrepreneurs. As mentioned in Chapters 4 and 5, they can become members of the extended family in the distributed innovation system.

Developing Project Leaders in an Entrepreneurial Perspective

A major bottleneck in technology companies is the lack of high-performance project managers. These are best developed by tackling a variety of challenges in order to develop their ability for getting things done through people, leading and motivating project members. Project leaders must be well equipped with ways of effectively managing across cultures, as well as having the following skills:

Project management
Innovation and research are very much about new ideas and change. Project leaders must develop their project management skills, including tools such as planning, organizing and executing the project in a business perspective. These tools include budgeting and planning, PERT and Gant charts, appropriate management software, as well as mastering the preparation of an effective *business plan*. The project leader must know the business language and be able to tightly link the technical development and its business impact for the firm. It is critical for project managers to have a good understanding of intellectual property management and of patent-based in- and out-licensing. As mentioned earlier, the effective development of project managers must be a top priority in technology firms.

Finance management
There is often a gap in the training of technical personnel. Training employees in managing personal finances, including personal investments, tax legislation and retirement funds, is a good place to start. Such courses provide insights into financial matters, which can be put directly to use in the course of a professional activity. They also give employees a tool for improving their own personal financial situation. Although this practice seems to make eminent sense, it is not at all widespread in companies. Elements of corporate finance are also part of this training. An understanding of the mechanics of the venture capital industry should also be included.

Short-term expatriation
Carefully prepared 'internships', lasting a few weeks, in a different part of the firm's operations constitute a powerful management development tool. By spending three well-prepared weeks in a plant, a marketing professional learns a lot about the world of manufacturing in his or her

firm. It is surprising that such short-term expatriation for management development is not used more by companies. Internships seem to be seen as only relevant to students, who do it as part of their education. They should be used in the course of professional life as well: by spending a well prepared few weeks in a plant or in a marketing department, an engineer or scientist will have a tremendous learning experience.

The Richness of Diversity in a Team

Our world is increasingly interdependent. Thankfully, it is not homogeneous. The richness of its diversity must be appreciated and taken into account. A good way to do this is for managers and project team members to master several languages, including studies of so called 'dead' languages, such as Latin and Greek. Knowing languages goes well beyond being able to converse in it; it also provides considerable enrichment by allowing a person to see the world from different *points of view*.

In Europe in particular, professionals have the advantage of having to speak at least two or three languages, one of which is likely to be English. People having English as their mother tongue may, however, well be at a disadvantage, since they too easily come to believe that they do not need to learn another language. As a result, they miss out on having another *perspective*, which enables better multi-cultural management. In any unit, what should absolutely be avoided is to have a collection of clones – all from the same country, same school, same discipline and trajectory, comfortable with each other, but devoid of creative tension. On the contrary, a highly diverse staff, coming from different national backgrounds, fosters a healthy debate between individuals representing different viewpoints, while also providing access to widely different external networks. An Italian physicist brings his or her own specific set of contacts with Italian laboratories, universities, scientific journals and media, while a Chinese colleague brings a whole different perspective, experience and network. Staff diversity makes it impossible to have, literally, a single *point of view*. Staff diversity is highly conducive to innovation, as it promotes debate and dialogue.

If properly managed, this diversity, when perceived as a positive asset by management, will result in enhanced innovative spirit, energy and motivation of a development unit, as illustrated by the following examples.

In the technical services organization Battelle, which counts 7000 employees in North America and Europe and is headquartered in Columbus, Ohio, a study was carried out to identify its most innovative units over a period of five years. The measure was the number of patents granted to a unit over the period, as well as their potential value. Since patent filing alone is a questionable measure of innovation, it was qualified by an evaluation of the strength and effective value creation of the patents. The study showed that the top performing units were in one of the European laboratories, and the reason for this was the highly diverse multi-cultural/multinational make-up of their professionals.

Similarly, the most productive unit – in terms of innovations per capita – in the pharmaceutical R&D system of the then Glaxo-Wellcome company was, for a long time, its laboratory in Geneva, Switzerland. Again, the key reason for this was the highly multinational character of its employees who, with their families, were attracted from many different countries to come and live in the cosmopolitan and urbane city of Geneva.

Practising distributed innovation draws on a wide range of diverse organizations in various cultures of the world. These organizations may be universities, small private firms or start-up companies, as well as public R&D units. Managers and leaders involved in multi-participant projects have to bridge cultural divides between organizations. They will need to appreciate and benefit from the great value represented by the multiple points of view provided by the richness of diversity.

Conclusion

In successfully identifying and developing innovations, the talent and motivation of the staff members involved are absolutely crucial. Their high motivation constitutes a tremendous lever in achieving effective innovation. Considering this, it is another paradox that companies are not more mobilized and proactive in dealing with these issues, by ensuring the appropriate environment, as well as by selecting and developing their staff.

A supportive and 'walk around' style is what is required from the management supervising innovation projects. Much attention and care must be dedicated to developing the staff involved in these projects. In this regard, fostering a sophisticated business sense in an entrepreneurial perspective is particularly critical in successfully generating the powerful value creation of distributed innovation.

CHAPTER 8

Conclusion: Creating Value and Growth through Distributed Innovation

It is pointing out the obvious to say that we are living in a turn-around world. Business has responded to some of the changes with extensive restructuring. When it comes to technical innovation developments, however, firms have been more cautious in modifying their mode of operating. It is high time for technology firms to envisage innovation in a new perspective.

Although 'innovation is the key to growth' is a universally accepted phrase, CEOs are not truly committed to making sure that the innovation engine works effectively. This paradox must be resolved by having their environment encourage them to become true champions of innovation.

One powerful response to the new reality is *distributed innovation*. This novel approach involves widely expanding the firm's innovation perimeter in order to raise new revenues and graft external inputs onto innovation projects presenting high growth and value-creating potential. For this to work, CEOs must both champion these high-risk, high-reward ventures and also pay careful attention to conditions conducive to motivating human talent and to creating a culture of innovation, entrepreneurship and trust.

A Turnaround World

Both leaders of the remarkable metamorphoses of Nokia and Samsung Electronics, discussed earlier, saw the problems caused by the oil crisis of the 1970s as a triggering event, which prompted them to transform their companies. Indeed, that crisis marked the beginning of a remarkable period. The last thirty years have seen an unparalleled increase in the purchasing power of the citizens of the first world. Communication and travel have become much more accessible to all. The increase in trade has been spectacular.

There is, however, a less positive side to the picture. There are huge disparities between the first world and the developing world, in demographics as well as in living conditions, including the ravages of the Aids pandemic. No doubt this imbalance will continue to be a major geopolitical concern in our volatile and uncertain world.

Science and technology have been crucially shaping this period. Not surprisingly, the world of technical developments has also changed considerably. Technology flows have grown exponentially. This is partly the result of people's mobility: technology is best transferred by the participants themselves moving from one firm to another, from the university to a company R&D laboratory. One proof of the increasing trade in technology is that in less than ten years, licensing royalties multiplied by more than a factor of seven to reach $142 billion worldwide in 2000. The United States, the European Union and Japan account for more than 90 per cent of this trade. Another indication of the increasing flows of technology during that period was the development of contract research, brought to Europe by the US-based Battelle Institute in the 1950s. Government laboratories in the USA, Harwell in the UK, as well as universities and private firms have been increasingly selling innovation projects to help their clients become more competitive.

In the same period, Japan bootstrapped itself into the position as the second economy in the world. It did so largely by effectively leveraging technical developments. As part of this focused effort, Japan intrigued the world with the extensive cooperation taking place among competing technology companies.

As our world has become more and more interdependent, so technology flows have become more massive and extensive. Sources of technologies have grown in number and become more accessible.

People began to talk about networked technical innovations, just as the ICT – Information and Computer Technologies – were permanently impacting all activities. The development of software such as Linux experimented with a totally open approach of 'community development', in which everybody was able to contribute via the Internet.

All this primarily concerns the first world, where 90 per cent of the world's R&D investments are made. Paralleling the negatives mentioned above is the fact that the already large 'knowledge gap' is further widening. The dynamic world's factory of China, however, is forging ahead and is soon likely to become a science and technology powerhouse as well. Locating R&D facilities in China is fast becoming a must for technology companies. They will thus not only tap into a vital science and technology scene, but also participate in new management approaches that will no doubt emerge in that country.

During this period, companies have responded to the changing world by engaging in massive restructurings of their businesses. We saw that in some rare cases, such as Nokia and Samsung, this transi tion produced totally new companies. The restructuring took place at a national level as well: for example, Japan essentially shut down its aluminum production industry to import this metal from Australia and other places instead.

No such radical changes have affected the way technology companies envisage their innovation developments. Still, some reorganization took place: centralized laboratories were dismantled because they were not directly sufficiently relevant to the business any more. Considerable business sense was injected into the outlook of technical personnel. New tools and practices have been introduced. Among them is the multi-functional approach for innovation projects. Innovation projects become much more tightly linked with the business strategy of the company. Corporations such as ABB, IBM and 3M invested heavily in connecting their multiple development sites around the world. The process of innovation, however, was kept mainly internal to the firm, with occasional formalized inputs coming from outside, but on an uncoordinated and *ad hoc* basis.

This relatively slow adaptation of the innovation system comes from a number of factors. First, technology firms have a protective attitude towards the 'strategic' element represented by the know-how generated by their R&D investments over the years. Technology firms choose to believe that they have accumulated know-how to form an effective defence against competition, and that importing more from the outside would be a waste. There is a fear that engaging in more contact with the outside world might endanger that position. Second, it is difficult to evaluate the investments in R&D *per se*; it is a challenge to assess the return on such investments, even with hindsight. This lack of clarity makes it difficult to compare the merits of internal development versus external innovations.

Innovation is the Key to Long-term Growth of the Business

The lesson from this period is that innovation is the key to the future. The developments during the 1990s in the USA seem particularly convincing. The remarkable continuous economic growth during that decade was largely attributed to technical innovation. ICT was thought to be a key contribution to this golden era. As a result, US media then took a part for the whole when talking about 'technology stocks' to specifically refer to ICT companies, seeming to ignore that these do not by any means constitute the whole technology-intensive industry.

A strong consensus has thus emerged that innovation and R&D investments play a key role in creating value and wealth in our firms. The stakeholders of our industrialized societies are trumpeting slogans to the effect that today's R&D investments create the jobs for tomorrow. Given this unanimous opinion, one would expect that making the technical innovation process work effectively would constitute a high priority on the CEOs' agenda. This is not the case. Instead, short-term financial performance dominates this agenda, at least in the conventional, publicly traded companies.

Resolving this paradox starts by creating conditions which will bring about the necessary change of priorities. This requires that the criteria for the choice of the CEO must include long-term growth through innovation. Further, more 'innovation activists' on the boards of companies are needed, as well as among the most influential shareholders. The rewards systems of top management will have to become consistent with this goal. This includes adjusting the remuneration of CEOs to how well the company is doing, as compared with the industry average.

On all these issues, informed and engaged shareholders must become a more positive force. Without this, companies may elect to avoid the excessive tyranny of the short term by becoming private, as has been observed already in certain countries.

Resolving the Paradox through Distributed Innovation

Existing ways of envisaging technical innovation have not kept pace with the dramatic changes outlined above. A primarily internally focused innovation process is too constraining. External contributors must be engaged much more proactively. Technology firms have to considerably extend their *innovation perimeter*. They must 'seamlessly' associate internal and external actors in their innovation process.

By considerably enlarging the innovation perimeter, distributed innovation works in two ways. First, it creates new revenues (this is discussed in Chapter 5). Second, (discussed in Chapter 6) after proactively identifying high impact innovations, it makes it possible to marshall considerable external technologies, so that the firm is able to access additional options and may vastly enhance its own innovation development capabilities.

Boost Revenues from Technology

A number of channels must be better utilized in order to exploit technical expertise to the full. They will make it possible to increase substantially the return on R&D investments. These are as follows:

- *Licensing income*. Following IBM's example of dramatically tripling its licensing revenues within three years, evaluate your company's pool of intellectual property and know-how in search for items on which licensing contracts could be developed. Constantly scout for possible 'customers' of your technology.

If a firm has 'patents dormant in a drawer', it has idle and costly assets, as the yearly patent maintenance fees can add up to important amounts. At the very least, the exercise of looking at the market will allow you to assess the value of your patent portfolio and eliminate the dead wood.

- *Sell innovation projects*. Extract value from the innovation projects which have been discontinued. In such projects, certain elements might still be of high value to other firms. Potential buyers are likely to be companies you know well: suppliers or customers. Instead of fully sinking the cost of such projects, do as Philips did with its laser development. The selling effort must be able to rely on an excellent knowledge of the technology scene. It must start very soon after the decision to discontinue a project, since most likely its 'shelf-life' is short.

Innovation mining. In certain cases, a piece of your technology puzzle will create much more value if it is associated with another complementary piece, that you identify in another organization – one company or other. By combining these complementary pieces, you may be able to multiply the value of the joint package, either to be further co-developed or sold.

Venturing. This route involves a sustained effort to unlock the value of technology. Indeed, spinning out an innovation project into a separate company requires extensive management time, nurturing and staff development. This route may be the most effective to create value from the portfolio of available options. In successful cases rewards may be handsome, as seen in the example of Cambridge-based Generics.

Apply Distributed Innovation for Profit and Long-term Growth

In moving away from 'short-termism', the firm regularly identifies 'high impact' products and services that have the best potential to contribute in value creation. In developing offerings to meet these opportunities, draw extensively on external technical input to complement your own. Such longer-term innovation projects are carried out in an entrepreneurial perspective: the firm carefully selects the opportunities, then marshals the technical resources to address them.

This outsourcing of technology must not be done in an opportunistic manner or an *ad hoc* basis, as it has been done until now. What is needed is to *outsource with a clear purpose* aimed at these 'high impact' offerings. This involves an integrated effort to source the external technology to be incorporated in the development of solutions of the selected targeted opportunities. Marshalling the various inputs, external and internal to the firm, is aimed to provide the best possible technical toolkit towards ensuring the commercial success of the 'high impact' innovation projects. From this perspective, shopping for technology is done with the clear objective of achieving *effectiveness* and *short time to profit* for the innovation projects considered.

The inputs of external technology flow through the same channels as those discussed above. License-in technology from third parties. Buy innovation projects from other firms or laboratories. Buy appropriate start-ups in order to acquire their valuable technology. An example is Synaptics buying the Absolute Sensors start-up to access superior sensor technology. Do as Cisco did, and integrate them at high speed. You will thus rapidly enhance your ability to deploy solutions to address opportunities you have selected. Activate additional external channels as well. These include university laboratories – be curious about their research to stimulate your own – as well as relevant contract research organizations; monitor their activities to see where they can contribute, as they might bring other pieces to complete your technology puzzle.

Of all industrial sectors, the pharmaceutical industry is using such technology channels the most. One reason is that the industry is heavily science-based; it has to remain closely connected with university research. It must also purchase molecules to strengthen its pipeline of drug developments, often low on potential 'blockbusters'. Finally, pharmaceutical companies have the financial muscle to buy external technology, whether licensing, funding contract research, or investing in start-ups to have a window on their development. They are expected soon to practise fully distributed innovation through closely integrating external inputs. They will thus play with both hands on the keyboard, instead of picking at tunes with one finger.

The Way Forward

Today, very few companies truly practise distributed innovation. We have discussed the specific examples of Generics and Samsung, Cisco, Intel, Nokia and the pharmaceutical companies, which master important elements of it. There is no doubt that technology companies will increasingly apply this model in the future. The reason for this is simply that, alongside their internal innovation process, firms must leverage external technical expertise better if they want commercial success. This 'high risk, high reward' approach provides new options for effective value creation and growth.

Distributed innovation truly takes into consideration the fact that there is much more going on outside the firm than there is inside. For these occasional, carefully selected projects, the company acts as an *integrator* of a great diversity of technical sources.

The entrepreneurial perspective typical of this approach is consistent with the fact that technology companies need to have more *researchers–entrepreneurs*. The venture capital (VC) industry has existed since 1946, when Georges Doriot founded his company ARD – American Research Development, in Boston. Since that time, the world of corporate technical innovation has been surprisingly impervious to the discipline of value creation, a characteristic of the VC industry. The VC perspective, applied to the 'high impact' projects, is fully consistent with our objective of strong business growth over the longer term. With it comes the useful notion of *due diligence* both for evaluating innovation projects and assessing external technical input. It is time to inject such a perspective into the way technology companies approach the innovation process today.

With distributed innovation, the company R&D function increasingly acts as a *broker* of technology. This implies a schizophrenic dimension, since technical professionals working on internal developments may recommend that their firm buy a technology that might make their own obsolete. This creates yet another tension in management. In making distributed innovation work, it is particularly crucial that top management handle the human factor with great care, not only to maintain the conditions for a high motivation level, but also to enable a strong climate of trust.

Project leaders of ventures involving distributed innovation must be carefully developed for their complex job. As they deal with a very diverse set of cultures and organizations, they must thrive on turning the *richness of diversity* into a critical asset for the commercial success of their endeavours.

Because it relies so much on scanning and evaluation of the external environment, distributed innovation will greatly reinforce the outward perspectives of the staff in the firm. It will also constitute a great stimulus for *learning*. This is further reinforced by delegating staff to partner firms for extended periods of time, or by carefully organizing *internships* in different parts of the firm, as well as in other companies.

Above all, distributed innovation is the way for the CEO to facilitate a process enhancing the value creation for the firm. It is the key to resolving the innovation paradox. Moving in this direction will increasingly make technology companies *architects* of innovation: they secure the most appropriate elements to achieve the attractive design they have come up with. In this way, technology companies define their groundbreaking products and services with fewer development constraints than if they mainly rely on what they can do in-house.

Companies will continue to need the strong technical expertise of the internal R&D function. First, it has a key role in innovation projects, those leveraging distributed innovation, as well as those with a more internal focus. Second, a strong R&D function is needed to enable the firm to be an effective scout and buyer of external technology. That function will be somewhat smaller in size and much more outward looking that is presently the case.

By extensively opening their innovation system to external contributors, technology companies will unleash new potential for growth and job creation. They will achieve this by more effectively converting their large pool of existing technical knowledge into economic value. Our world needs such firms to contribute healthy value and job creation through a high and sustained rate of growth.

Abegglen, J. and G. Stalk (1985) *Kaisha: The Japanese Corporation*. Tuttle Company.

Allen, T. (1977) *Managing the Flow of Technology*. MIT Press.

Baumol, William J. (2002) *The Free-Market Innovation Machine*: Analyzing the Growth *Miracle of Capitalism*. Princeton University Press.

Bebear, Claude (2003) *Ils Vont Tuer le Capitalisme*. Plon.

Brockhoff, K. (1990) *Management von Forschung, Entwicklung und Innovation*. Stuttgart, Metzlersche Verlagsbuchhandlung und Carl Ernst Poeschel Verlag GmbH.

Brooklyn, Den C. (1988) *Managing the New Careerists*. Jossey-Bass Publishers.

Bush, Vannevar (July, 1945) *The Endless Frontier*. United States Government Printing Office.

Cambridge Phenomenon, The: The Growth of High Technology Industry in a University Town (1985) Segal Quince Wicksteed Ltd.

Chesbrough, Henry W. (2003) *Open Innovation: The New Imperative for Creating and Profiting from Technology*. Harvard Business School Press.

Christensen, Clayton M. (1997) *The Innovator's Dilemma: When New Technologies Cause Great Firms to Fail*. Harvard Business School Press.

Cooper, Robert G. (1993) *Winning at New Products: Accelerating the Process from Idea to Launch*. Corporate & Professional Publishing Group.

Deschamps, J-P and Nayak, P. R. (1995) *Product Juggernauts: How Companies Mobilize to Generate a Stream of Market Winners*. Harvard Business School Press.

Dosi, Giovanni, Christopher Freeman, Richard Nelson, Gerald Silverberg and Luc Soete (1988) *Technical Change and Economic Theory*. Pinter Publishers Ltd.

Dosi, Giovanni, David J. Teece and Josef Chytry (1998) *Technology, Organization, and Competitiveness: Perspectives on Industrial and Corporate Change*. Oxford University Press.

Drucker, Peter F. (1985) *Innovation and Entrepreneurship*. Butterworth Heinemann.

Foster, R. and S. Kaplan (2001) *Creative Destruction: Why Companies That are Built to Last Underperform the Market – and How to Successfully Transform Them*. Doubleday.

Fransman, Martin (1999) *Visions of Innovation: The Firm and Japan*. Oxford University Press.

Frascati Manual 1993, Proposed Standard Practice for Surveys of Research and Experimental Development (1994) OECD Publications.

Gaynor, Gerard H. (1996) *Handbook of Technology Management*. McGraw-Hill.

Grove, A. (1999) *Only the Paranoid Survive*. Random House.

Häikiö, Martti (2002) *Nokia: The Inside Story*. Pearson Education.

Haour, Georges and André P. Maïsseu (1995) Management of Technological Flows Across Industrial Boundaries, *International Journal of Technology Management*, Volume 10, No. 1. Inderscience Entreprises Ltd.

Hayashi, Takeshi (1990) *The Japanese Experience in Technology: From Transfer to Self-Reliance*. United Nations University Press.

Hermann, Simon (1996) *Hidden Champions: Lessons from 500 of the World's Best Unknown Companies*. Harvard Business School Press.

Hitt, William D. (1988) *The Leader-Manager: Guidelines for Action*. Battelle Press.

Howells, Jeremy and Jonathan Michie (1997) *Technology Innovation and Competitiveness*, Edward Elgar Publishing Limited.

Gillies, James and Robert Cailliau (2000) *How the Web Was Born*. Oxford University Press Inc.

Jolly, Vijay K. (1997) *Commercializing New Technologies: Getting from Mind to Market*. Harvard Business School Press.

Kealey, Terence (1996) *The Economic Laws of Scientific Research*. Macmillan Press Ltd, London and St Martin's Press Inc.

Koestler, Arthur (1989) *The Act of Creation*. Arkana.

Kuhn, Thomas S. (1962) *The Structure of Scientific Revolutions*. International Encyclopedia of Unified Science. University of Chicago.

Le Duff, Robert (1999) *Encyclopédie de la Gestion et du Management*. Dalloz, Paris.

Low, Morris, Shigeru Nakayama, Hitoshi Yoshioka (1999) *Science, Technology and Society in Contemporary Japan*. Cambridge University Press.

Matsushita, Konosuke (1984) *Not for Bread Alone: a Business Ethos, a Management Ethic*. PHP Institute Inc.

Meurling, John and Jeans, Richard (1994) *The Mobile Phone Book: The Invention of the Mobile Telephone Industry*. Communications Week International.

Michie, Jonathan and John Grieve Smith (1998) *Globalization, Growth, and Governance*. Oxford University Press Inc.

Morin, Jacques (1985) *L'Excellence Technologique*. Publi-Union.

Morita, Akio (1986) *Made in Japan*. William Collins Sons & Co. Ltd.

Nathan, John (1999) *Sony: The Private Life*. HarperCollins Business.

New Economy, The: Beyond the Hype, The OECD Growth Project (2001) OECD Publications.

Nonaka, Ikujiro and Takeuchi, Hirotaka (1995) *The Knowledge-Creating Company: How Japanese Companies Create the Dynamics of Innovation*. Oxford University Press.

Peters, T. (1997) *The Circle of Innovation*. Coronet Books.

Rice, M. *et al.* (2001) *Radical Innovation: How Mature Companies Can Outsmart Upstarts*. Harvard Business School Press.

Schumpeter, Joseph A. (1911) *The Theory of Economic Development*. Harvard University Press.

Steinbock, Dan (2001) *The Nokia Revolution: The Story of an Extraordinary Company That Transformed an Industry*. AMACOM.

Tidd, Joe, John Bessant and Keith Pavitt (1997) *Managing Innovation: Integrating Technological, Market and Organizational Change*. John Wiley & Sons Ltd.

Tushman, Michael L. and Charles A. O'Reilly III (1997) *Winning Through Innovation: A Practical Guide to Leading Organizational Change and Renewal*. Harvard Business School Press.

Utterback, J. (1994) *Mastering the Dynamics of Innovation: How Companies Can Seize Opportunities in the Face of Technological Change*. Harvard Business School Press.

Von Braun, Christoph-Friedrich (1997) *The Innovation War: Industrial R&D ... the Arms Race of the 90s*. Prentice-Hall Inc.

Von Zedtwitz, Max, Georges Haour, Tarek M. Khalil and Louis A. Lefèbvre (2003) *Management of Technology: Growth through Business Innovation and Entrepreneurship*. Pergamon, Elsevier Science Ltd.

Weimer, William A. (1992) *Masters and Patrons: Renaissance Solutions for Today's Productivity Problems*. Dogwood Publishing Company.

WheelWright, Steven C. and Kim B. Clark (1992) *Revolutionizing Product Development: Quantum Leaps in Speed, Efficiency, and Quality*. The Free Press.

Index

Key: f=figure illustration; n=note; t=table; **bold**=extended discussion or heading emphasized in main text.